From Sunlight to Electricity: Exploring Nanocrystals for Solar Energy Conversion

Monty

First Printing, 2024

TABLE OF CONTENTS

1. INTRODUCTION

Lighting applications is one of the most interesting and important topic in human society, and many researchers are devoting themselves to find new materials and technologies in this area. There are variety of solutions starting from the well-known commercially available incandescent sources, light emission diodes (LED) lamps, fluorescent lamps, which produce white-light (WL). The first is not energy efficient while the latter two do not produce the black-body emission like the sun does that is most pleasing to the eye.

In the past decade, the generation of black-body WL emission from nanostructured materials doped and un-doped with lanthanides and excited with near infrared (NIR) laser radiation, has been a subject of different research studies. The WL emission is a combination of visible colors emitted by trivalent dopant rare earths ions (REI) such as Yb^{3+}, Tm^{3+}, Er^{3+}, Nd^{3+} [1-4] in the nano-phosphor host. These REI doped materials convert infrared radiation to visible light through an upconversion process, and are known as "optical transducers" [5]. These have a greatly promising application in displays, florescent lamps, phosphor-based lamps, temperature sensors etc.

In this book, I carry out a systematic study of WL generation from REI materials through the upconversion process in a nano-phosphor host. The emission from REIs has been studied in the past by various groups. Among the upconversion materials, the Er^{3+} and Yb^{3+} ions have excellent upconversion properties [6-8]. The emission of REIs is in the $4f^n$ energy levels which are well protected by the $5s^2$ and $5p^6$ electrons, so the host material does not impact the REI energy levels much. Er^{3+} has a strong absorption at about 520, 550 and 654 nm due to its $^4S_{3/2}$ - $^4I_{15/2}$, $^2H_{11/2}$ - $^4I_{15/2}$ and $^4F_{9/2}$ - $^4I_{15/2}$ transitions which produce green and red luminescence, while Yb^{3+} has its absorption at 1000 nm due to $^2F_{5/2}$ - $^2F_{7/2}$ transition which produces infrared. Co-doping with Yb^{3+} and Er^{3+} has proven to be a successful alternative for the upconversion process in lots of nanocrystalline [10-11] because of the large spectral overlap between Yb^{3+} NIR emission and Er^{3+} absorption bands, which result in an efficient energy transfer [11]. Moreover, the absorption wavelength of Yb^{3+} also matches with the emission wavelength of the commercially widely used NIR laser diodes (975 nm).

Nano sized crystalline systems have many advantages for the optical and electronic properties due to better thermal effects and higher luminescence efficiency compared to the bulk counterparts. The REIs doped nanocrystallines are widely used in photonics area, and the exploration for new, efficient and convenient materials has never been stopped. In 1994, a new nanocrystal phosphor of ZnS doped with Mn was found to produce high luminescence efficiency due to its confinement effect by Bhargava's group [12]. Since then, many researchers have started to investigate the optical properties of nanocrystal insulating semiconductors [12–15]. Some oxide nanocrystals (NC) doped with REIs are important phosphors for the luminescence application [16–19]. For instance, RE doped phosphors such as Gd_2O_3, $Y_4Zr_3O_{12}$, LaF_3 etc. are commonly used in optical display. There are immense amounts of research on REIs doped nanocrystal phosphors for luminescence and WL study in the literature. Among them, Er^{3+} and Yb^{3+} codoped cubic $SrZrO_3$ crystals have numerous excellent performance such as wide band gap, high permittivity, high thermal conductivity, high disruptive strength, low Leakage current density and good chemical and mechanical stability [20-21]. It has been used in fuel cells, hydrogen gas sensors, steam electrolysis, thermal barrier coatings and so on [20-21]. Therefore, people pay more and more attention to its experimental research [22-24] and theoretical calculation [25].

In our study, the synthesis work of the specimens $SrZr_{1-x}Er_yYb_xO_{3-0.5x}$ (x = 0, 1%, 2%, 4%, 8%; y = 0, 1%) was done by *Professor Federico Garcia Gonzalez* in *Universidad Autónoma Metropolitana-Iztapalapa* in Mexico. Totally we have nearly 40 samples including pellets and powders. The pellets were annealed at 2 different temperatures: 1200 °C and 1600 °C, while the powders were annealed at 900 °C and 1100 °C. For the same category, a higher annealing temperature gives a bigger nanoparticle size, and their orthorhombic perovskite-like structure was confirmed by X-ray diffraction.

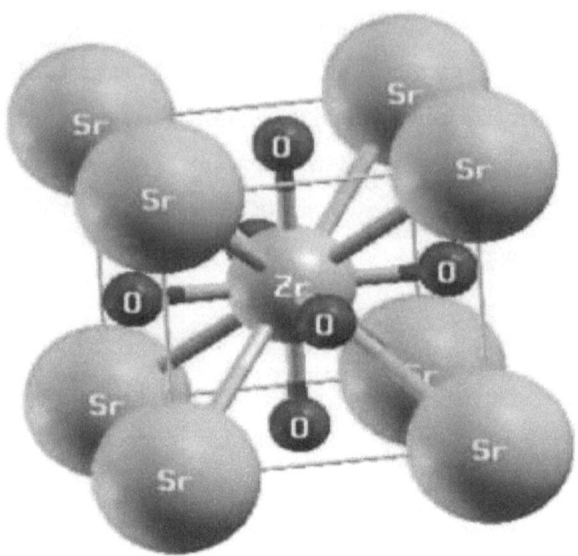

Figure 1. 1. Crystal structure of cubic $SrZrO_3$

Most of the research on Er^{3+} - Yb^{3+} doped systems has been studied well in the past years. The upconversion host materials like transition metal oxides [26-27], rare-earth oxides [26, 28-29], and zirconates [30-31] have been concentrated due to their good thermal, mechanical and chemical properties. However, there is not much detailed report for the luminescence and WL phenomena based on $SrZrO_3$ system. It is also a good candidate in optics study, because it has not only the good physicochemical properties as stated above, but also high melting point of 2600 °C and low maximum phonon energy of ~ 750 cm^{-1}, which favors the light producing. Therefore, the nanocrystalline $SrZrO_3$ doped with Er^{3+} - Yb^{3+} gives us a good chance to investigate for upconversion process and lighting applications.

In this work, we would like to focus on the generation of luminescence and WL in $SrZrO_3$ nanocrystalline doped and nominally un-doped with Er^{3+} - Yb^{3+} by a near-infrared (NIR) laser excitation. At low powers, luminescence is generated, while at high powers above the threshold, a broad emission spectrum ranging from 400 nm – 800 nm is obtained. Another important feature of the NIR excitation is that the excitation wavelength may be chosen freely from visible and NIR spectrum leading to efficient Stokes and anti-Stokes excitations.

The over years' experience in the WL emission investigations will be discussed giving a perspective look on this internally complex phenomenon. Special attention will be given to possible mechanisms responsible for the WL emission generation. Important aspects of application of the WL emission will also be discussed. Currently developed solutions include new types of white light sources, solar emission to electrical current conversion, high mobility light-driven vacuum devices/detectors and novel types of transistors opening the way for dynamic change of working characteristics.

Despite all this effort, some important questions still remain open, such as the origin of the WL; the effect of the crystallite size and the role played by some dopant ions on its generation, and if the phenomenon is intrinsic or not to nanostructured systems. However, with the numerous studies on this topic, we believe that these problems will be gradually solved as time goes on, and it may change the future technology bringing a new look at the light generation.

The outline of this book is as follows: In Chapter 2, we will talk about the theoretical background such as nanocrystallines, rare earth elements, upconversion process and physicochemical properties of $SrZrO_3$ crystal. In Chapter 3 and 4, we will provide the experimental procedures for the sample preparation, and the introduction of the apparatus. In Chapter 5 and 6, we will investigate the results of the production of luminescence and WL. Finally, in Chapter 7, we set the conclusions and the future prospects of the work.

2. BACKGROUND

2.1 NANO POWDERS

Nano-powder is also called nano-particle, generally refers to the size of ultrafine particles between 0.1-100 nm, some people call it ultrafine particles [32]. Its size is larger than that of clusters and smaller than that of general particles. In terms of its size, assuming that each atom is one angstrom in size, the number of atoms it contains ranges from 1,000 to 1 billion. The morphology of nanoparticles is spherical, plate, rod, horn, sponge and so on [33]. The components of nanoparticles can be metals, oxides and other compounds [34].

Currently, nanomaterials are defined in the range of 0.1-100 nanometers. When the particle size is less than 100 nanometers, the number of atoms on the surface of the particles is comparable to that in the body. For example, 5 nanoparticles have a surface atomic ratio of 40% and a specific surface area of 180 square meters per gram. This leads to the appearance of small size nanomaterials which are different from the traditional solid materials or their bulk counterparts, due to the change of the structure and energy state caused by the surface effects [35] and quantum tunnel effects [36]. Many unique physical and chemical energies, such as light, electricity, magnetism and mechanics, have been produced by the changes. For example, nano-iron has magnetic properties, the original non-conductive materials become conductive, special far-infrared radiation, strong ultraviolet reflection, strong catalysis, strong adsorption and so on [37-38].

Surface effects of nanoparticles

With the decreasing of particle size, the number of atoms on the surface increases rapidly, and accounts for a considerable proportion of the total atoms. For example, when the particle size is 10 nm, the proportion of surface atoms reaches 50%; when the size is 5 nm, the surface atoms increases to 80%; when the size is 1 nm, the surface atoms reaches 99%, and almost all atoms are in the surface state. The surface free energy and chemical bond of nanoparticles are greatly increased due to the large number of surface atoms, and the bond states are seriously mismatched. Many active centers appear, the steps and roughness of the surface increase, and the chemical valence of non-chemical equilibrium and non-integer coordination appears on the surface. This leads to the difference of the chemical properties of nanoparticles compared with the traditional chemical equilibrium system. Moreover, the surface atoms have a lack of adjacent atoms, so they have many suspended bonds with unsaturated properties, and are easy for other atoms to combine and stabilize. Therefore, they show great chemical and catalytic activities.

2.2 RARE EARTH

Rare Earth (RE) is a general term for 17 elements of lanthanides and lanthanum and cerium in the chemical periodic table. The first man who discovered rare earths was the Finnish chemist John Gadolin. In 1794, he separated the first rare earth "element" (alkaline, Y_2O_3) from a bituminous heavy ore. Because rare earth minerals were found in the 18th century, only a small amount of water-insoluble oxides could be produced by chemical methods. Historically, this oxide was called "soil", hence the name rare earth.

According to the electronic layer structure and physicochemical properties of rare earth elements, and their symbiosis in minerals and different ionic radii, different characteristics can be produced. In this work we focus on the optically active ions in the lanthanide series, which includes the following elements: cerium(Ce), praseodymium(Pr), neodymium(Nd), promethium(Pm), samarium(Sm), europium(Eu), gadolinium(Gd), terbium(Tb), dysprosium(Dy), holmium(Ho), erbium(Er), thulium(Tm), ytterbium(Yb).

These rare earth ions (REI) have the atomic numbers from 58 (Ce) to 70 (Yb), and they have similar atomic structure and ion radius (Re^{3+} ion radius 1.06×10^{-10} m ~ 0.84×10^{-10} m, Y^{3+} 0.89×10^{-10} m), and are closely symbiotic in nature. The rare earth ions are commonly used for their luminescence properties, and have many applications in phosphors, lasers, superconductivity, etc.

In recent years, rare earth-doped compounds, as the treasure house of optical high-tech materials, have attracted more and more attention in their value and application. The trivalent state of rare earth elements is the characteristic oxidation state of REIs, and they have $4f^n$ electronic configuration with n values from 1 to 13. There are seven subshells in the 4f shell, with each subshell has different orbital angular momentum quantum number, l ($-3 < l < 3$). Transitions within the 4f configuration produce a rich fluorescence spectrum. Figure 2.1 shows the number of 4f electrons in the different trivalent REIs. The luminescence of rare earth ions has many excellent properties, which makes the study of the luminescence of rare earth elements have important theoretical significance [22-25] and application value [16–19].

Ion	n
Ce^{3+}	1
Pr^{3+}	2
Nd^{3+}	3
Pm^{3+}	4
Sm^{3+}	5
Eu^{3+}	6
Gd^{3+}	7
Tb^{3+}	8
Dy^{3+}	9
Ho^{3+}	10
Er^{3+}	11
Tm^{3+}	12
Yb^{3+}	13

Figure 2. 1. The number of 4f electrons in trivalent REIs

Rare earth ions have plenty of emission spectra, as there are up to 30000 energy levels within the 4f shell for all the elements combined. Therefore, they can emit electron radiation of various wavelengths from ultraviolet to infrared. Figure 2.2 shows the energy-level diagram for trivalent lanthanide rare earth ions in lanthanum chloride. The absorption spectral lines for these transitions within the 4f shell are very narrow, so it presents bright and pure color. The fluorescence spectra of rare earth ions are different from that of ordinary ones, and has a large Stokes displacement. For example, the Stokes displacement of $Dy(PTA)_3$ reaches 278 nm, and the emission spectrum and excitation spectrum will not overlap each other.

However, the Stokes shift of ordinary fluorescent materials is small, for example, the Stokes shift of fluorescein is only 28 nm, so the excitation spectrum and emission spectrum are usually partially overlapped, and the mutual interference is serious.

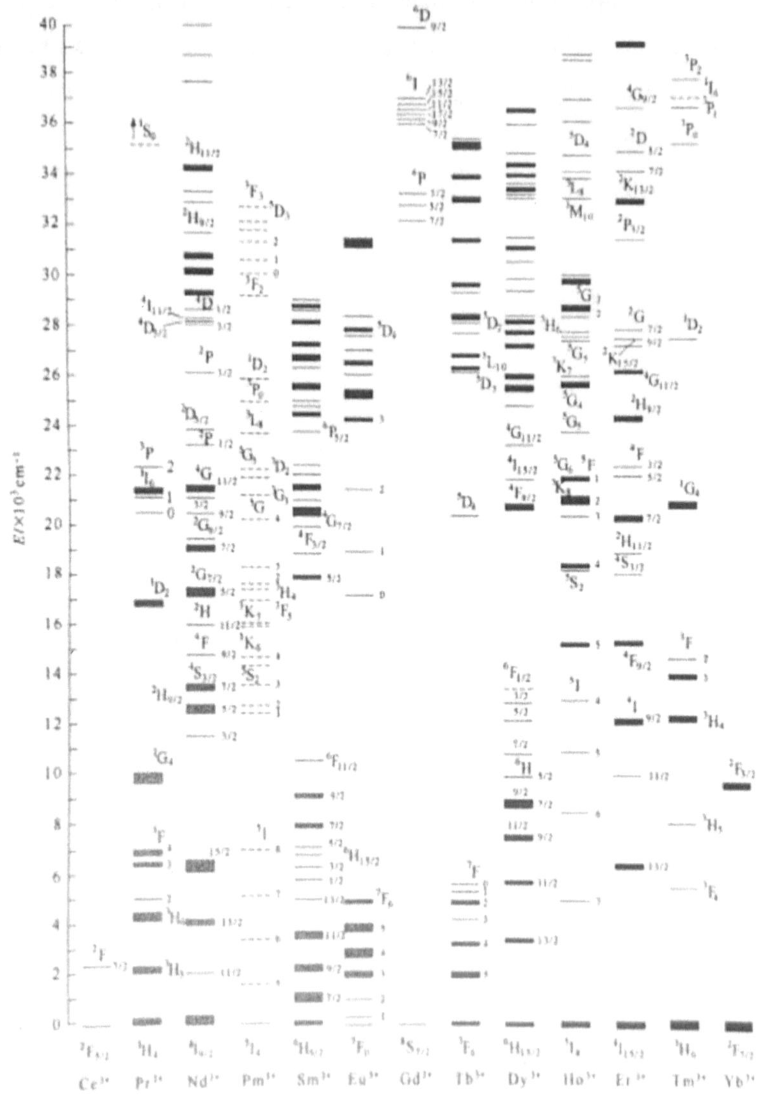

Figure 2. 2. An energy-level diagram for trivalent lanthanide rare earth ions in lanthanum chloride

2.3 BLACK BODY RADIATION

Each object has the property of continuously radiating, absorbing and reflecting electromagnetic wave. The radiated electromagnetic wave is different in each band, that is to say, it has a certain spectral distribution. This spectral distribution is related to the characteristics of the object itself and the temperature, so it is called thermal radiation. Blackbody refers to the object that completely absorbs the external radiation of any wavelength without any reflection under any conditions, that is, the object with an absorption ratio of 1. The ideal blackbody can absorb all the electromagnetic radiation on its surface, and convert this radiation into thermal radiation. Its spectral characteristics are only related to the temperature of the blackbody, but not to the material.

Planck's radiation law gives the specific spectrum distribution of blackbody radiation [39]. At a certain temperature, the energy radiated by blackbody per unit area in unit time, unit solid angle and unit wavelength interval is

$$B(\lambda, T) = \frac{2hc^2}{\lambda^5} \frac{1}{e^{\frac{hc}{\lambda KT}} - 1}$$

B (λ, T)--Spectral radiance of blackbody$(W \cdot m^{-2} \cdot Sr^{-1} \cdot \mu m^{-1})$

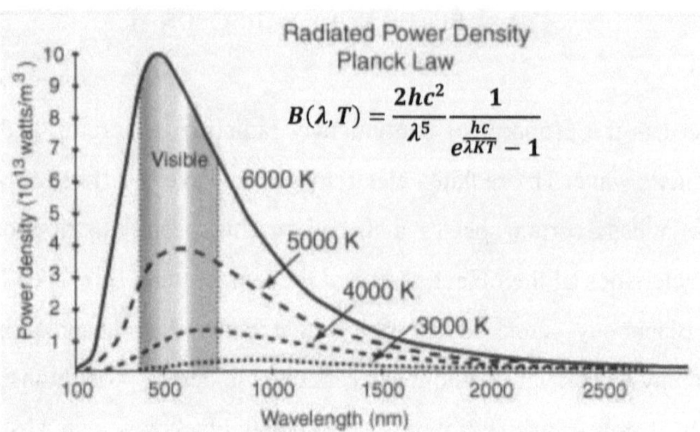

Figure 2. 3. Radiated Power Density by Planck Law

We can see from the figure above:

At a certain temperature, there is an extreme value of the spectral radiance of blackbody. The position of the extreme value is related to the temperature. This is Wien's law of displacement -- in the blackbody radiation spectrum, the product of the wavelength generating the maximum radiation and the temperature of blackbody is a constant:

$$\lambda_m \cdot \mathrm{T} = \mathrm{b}$$

b=0.002897m·K is called Wien constant

According to Wien's displacement law, we can estimate that:

1, when T = 6000K, λ_m = 500 nm (green). This is the approximate maximum spectral radiance of solar radiation.

When T = 300K, λ_m = 9.6 μm, which is the approximate maximum spectral radiance of the earth object radiation.

2, At any wavelength, the spectral radiance of the high-temperature blackbody is absolutely greater than that of the low-temperature blackbody, regardless of whether this wavelength is at the maximum spectral radiance.

If we integrate B (λ, T) for all wavelengths and also for all radiation directions, then Stefan Boltzmann law can be obtained. The total energy radiated by the unit area of blackbody with absolute temperature of T in each direction of space in unit time is B (T):

$$B(T) = \sigma T^4 \ (W/m^2)$$

where σ is Stefan Boltzmann constant, equal to $5.67 \times 10^{-8}\, W/(m^2k^4)$.

2.4 PHOTON UPCONVERSION

Photon upconversion, i.e. Anti-Stokes luminescence, is derived from Stokes' law. Stokes' law holds that materials can be excited by photons with high energy and emit photons with low energy. In other words, light with long wavelength and low frequency will follow from excitation with light having high frequency and short wavelength. In fluorescent lamps, for example, ultraviolet light (wavelength = 254 nm) is used to excite a phosphor, which emits in the visible region of the spectrum. However, it was found that some materials can achieve exactly the opposite effect, so we call it anti-Stokes luminescence, also known as up-conversion luminescence.

Up-conversion luminescence has occurred in REIs doped compounds, mainly fluorides, oxides, sulfur compounds, fluorides, halides and so on. $NaYF_4$ is the most efficient up conversion luminescence matrix material at present, such as $NaYF_4$: Er^{3+}-Yb^{3+} co-doped nanopowder [40-42]. When Yb^{3+} and Er^{3+} are doped, Er^{3+} acts as activator and Yb^{3+} acts as sensitizer.

Unlike the traditional typical luminescence process, which involves only one excited state and one ground state, the upconversion process requires at least one intermediate state. An ion, once excited in the intermediate state, may then absorb another low-frequency photon and be raised to a higher energy level. From that level, the ion may then emit a photon with a higher energy than each of the absorbed photons. There are three main upconversion mechanisms: excitation state absorption, energy transfer upconversion and photon avalanche [43-46]. These processes are accomplished by continuous absorption of one or more photons by doped active ions in a host. High efficiency up-conversion can be easily achieved with many of the trivalent rare earth ions [47-49] because of their long metastable state lifetime.

The upconversion process mainly depends on the several states of doped REIs. The absorption and emission spectra of REIs mainly come from the transition of 4f electrons in the inner layer. Shielded by the 5s and 5p electrons, the 4f electrons hardly interact with the matrix, so the absorption and emission spectra of doped REIs are similar to those of their free ions, showing very sharp peaks (half-peak width is typically 10-20 nm in the visible region at room temperature). Therefore, the wavelength of the excitation source is greatly limited to that narrow wavelength range.

Lanthanide metal ions usually have a series of sharp emission peaks, which provides a characteristic spectrum for spectral analysis, and avoids the influence of overlapping emission peaks. The emission peak wavelength is not affected significantly by the chemical composition and physical size of the matrix at all. By adjusting the composition and concentration of doped ions, the relative intensity of different emission peaks can be controlled, so as to achieve the purpose of controlling the luminescence color.

Unlike anti-Stokes processes that occur at the site of a single ion, such as two-photon absorption and excited state absorption, the upconversion process of interest for this work requires energy transfer between two different rare earth ions. In our case, the upconversion process was observed using continuous-wave diode laser in the infrared, which resulted in emission in the visible.

Because the upconversion luminescence process of the inner 4f electron transition does not involve the breaking of chemical bonds, upconversion nanoparticles have high stability without photo fading and photochemistry fading. Many independent studies have shown that REIs doped nanoparticles have not changed fundamentally after several hours of ultraviolet and infrared laser irradiation [50-51].

Er^{3+}-Yb^{3+} Upconversion

Er^{3+} emission spectra consist of three emission bands, the central wavelength of red emission is 650 nm, and the emission maxima in the green are 520 and 550 nm. These emissions correspond to the $^4F_{9/2}$- $^4I_{15/2}$, $^2H_{11/2}$- $^4I_{15/2}$, and $^4S_{3/2}$- $^4I_{15/2}$ transitions of Er^{3+} ions [10-11, 19, 26-29], respectively.

Up-conversion luminescence of Er^{3+}, Yb^{3+} ion-doped single crystals or glasses under 980 nm excitation has been extensively studied [10-11, 26-29]. The main processes are the absorption of excited states and the energy transfer from Yb^{3+} to the Er^{3+} ion. The upconversion process can be described by the following three steps:

1. The Yb^{3+} ion absorbs a photon from the infrared laser, causing a transition from its ground $^2F_{7/2}$ state to the excited $^2F_{5/2}$ state. (See Figure 2.4.)

2. If an Er^{3+} ion is nearby to the excited Yb^{3+} ion, then the Yb^{3+} ion can transfer its energy nonradiatvely to the Er^{3+} ion. This can occur because the Er^{3+} ion has an energy level ($^4I_{11/2}$) at an energy very close to the $^2F_{5/2}$ level of Yb^{3+}. (see Figure 2.4.) Thus, the energy transfer can conserve energy, which is required for the transfer to occur. This energy transfer will only be efficient when the doping levels are high enough so that the Yb^{3+}-Er^{3+} distance is not too large.

3. If there is another excited Yb^{3+} also nearby to the Er^{3+} ion, then a second Yb^{3+} to Er^{3+} nonradiative energy transfer can result in the Er^{3+} ion being excited from its excited to a higher energy level. The Er^{3+} ion may then emit in the visible region.

When the doping concentration is high, the energy transfers between the two ions, and resulting in upconversion luminescence, can successfully compete with other decay processes, and so the upconversion luminescence is easily observed in the laboratory.

As mentioned, we see three main emission peaks from the Er^{3+} ion at 520, 550, and 650 nm. Each emission results from different upconversion pathways, which are now described in detail. The reader is referred to Figure 2.4 for these discussions.

For the emission at 520 and 550 nm, the process is as follows. Yb^{3+} ions in the excited $^2F_{7/2}$ state transfers its energy to $^4I_{11/2}$ level of Er^{3+}. A second Yb^{3+} to Er^{3+} energy transfer causes the Er^{3+} ions be excited from its $^4I_{11/2}$ level to the $^4F_{7/2}$ level. Form the $^4F_{7/2}$ level it quickly relaxes non-radiatively to the $^4S_{3/2}$ and $^2H_{11/2}$ metastable levels. Finally, Green light emission (520, 550 nm) is achieved by $^2H_{11/2}$ - $^4I_{15/2}$ and $^4S_{3/2}$ - $^4I_{15/2}$ transitions [11, 27, 54]. Song et al. expressed the processes for population the green emissions as follows [19]:

$$^2F_{5/2}(\mathrm{Yb}^{3+}) + {}^4I_{15/2}(\mathrm{Er}^{3+}) \rightarrow {}^2F_{7/2}(\mathrm{Yb}^{3+}) + {}^4I_{11/2}(\mathrm{Er}^{3+})$$

$$^2F_{5/2}(\mathrm{Yb}^{3+}) + {}^4I_{11/2}(\mathrm{Er}^{3+}) \rightarrow {}^2F_{7/2}(\mathrm{Yb}^{3+}) + {}^4F_{7/2}(\mathrm{Er}^{3+})$$

$$^4F_{7/2}(\mathrm{Er}^{3+}) \rightarrow {}^2H_{11/2}/\,{}^4S_{3/2}\,(\mathrm{Er}^{3+})$$

For the red emission 650nm, the process is as follows. Yb^{3+} ions in the excited $^2F_{7/2}$ state transfers its energy to $^4I_{11/2}$ level of Er^{3+}. From the Er^{3+} $^4I_{11/2}$ level, the system relaxes to the $^4I_{13/2}$ level. A second Yb^{3+} to Er^{3+} energy transfer causes the Er^{3+} ions be excited from its $^4I_{13/2}$ level to the $^4F_{9/2}$ level. Finally, red light emission (650 nm) is achieved by the $^4F_{9/2}$ - $^4I_{15/2}$ transition. Song et al. expressed the processes for population the red emission as equations as follows [19]:

$$^2F_{5/2}(\mathrm{Yb}^{3+}) + {}^4I_{15/2}(\mathrm{Er}^{3+}) \rightarrow {}^2F_{7/2}(\mathrm{Yb}^{3+}) + {}^4I_{11/2}(\mathrm{Er}^{3+})$$

$$^4I_{11/2}(\mathrm{Er}^{3+}) \rightarrow {}^4I_{13/2}(\mathrm{Er}^{3+})$$

$$^2F_{5/2}(\mathrm{Yb}^{3+}) + \, ^4I_{13/2}(\mathrm{Er}^{3+}) \rightarrow \, ^2F_{7/2}(\mathrm{Yb}^{3+}) + \, ^4F_{9/2}(\mathrm{Er}^{3+})$$

Figure 2.4 shows the diagram of Er^{3+}-Yb^{3+} energy level upconversion transition.

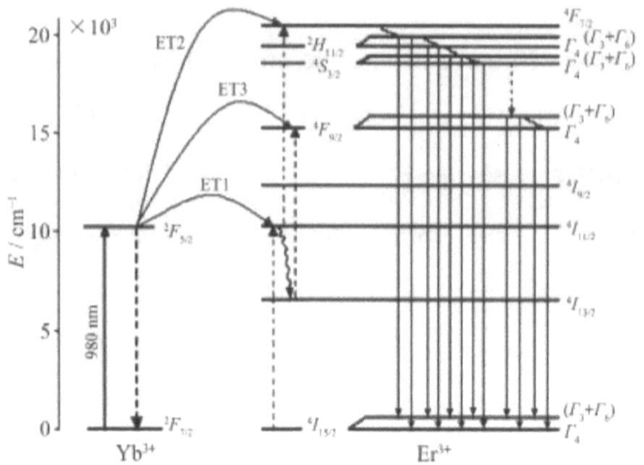

Figure 2. 4. Er^{3+}-Yb^{3+} Transition diagram

In the experiment, the emission bands of up-conversion emission spectra are composed of multiple spectral peaks. The emission bands near 525, 550 and 668 nm observed experimentally consist 3, 2 and 4 spectral lines near 525, 550 and 668 nm, respectively. The multiple lines associated with each energy level indicate that the energy levels of Er^{3+} are splits by the crystal field. This is evidence of the Stark effect.

In a free ion, the 4f electrons of the REIs are affected by a central force field of ions, and the spin-orbit interaction splits each spectral term is into corresponding energy level labeled $^{2S+1}L_J$ (J is the total angular momentum quantum number). When placed in a host, the REI also experiences a crystal field due to the effect of the surrounding ions. This crystal field further splits each $^{2S+1}L_J$ into Stark sub-levels. The number of Stark levels will depend on the local symmetry of the REI. In this way, at most several spectral lines of different wavelengths can be obtained for the transition between each excited state and ground state.

2.5 PROPERTIES OF THE $SrZrO_3$ CRYSTAL

$SrZrO_3$ (SZO) is a perovskite (ABO_3) oxide of orthorhombic structure with a 4.405 Å space group and a 85.497 $Å^3$ volume. It has a high melting point of 2600 °C and a low maximum phonon energy of ~ 750 cm^{-1}. Figure 2.5 gives the crystal structure of the SZO. SZO is a kind of solid electrolyte material as a high-temperature proton conductor, and of great interest with strong application prospect. It has many excellent properties, such as wide band gap, high dielectric constant, high breakdown strength, low leakage current density, excellent chemical and mechanical stability and so on. It is widely used in fuel cells, hydrogen gas sensors, dielectric materials, steam electrolysis, thermal barrier coating, etc. SZO also exhibits strong luminescence, making it a promising material for the optics applications.

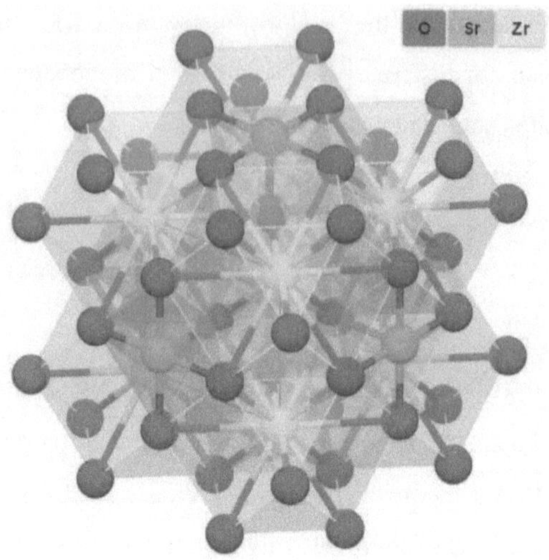

Figure 2. 5. Crystal structure of $SrZr0_3$

More information is calculated from the Materialsproject.org softwire: Figure 2.6 shows the essential parameters of $SrZr0_3$ structure, Figure 2.7 shows the electronic structure of $SrZr0_3$, and Figure 2.8 shows the X-ray diffraction patterns of $SrZr0_3$.

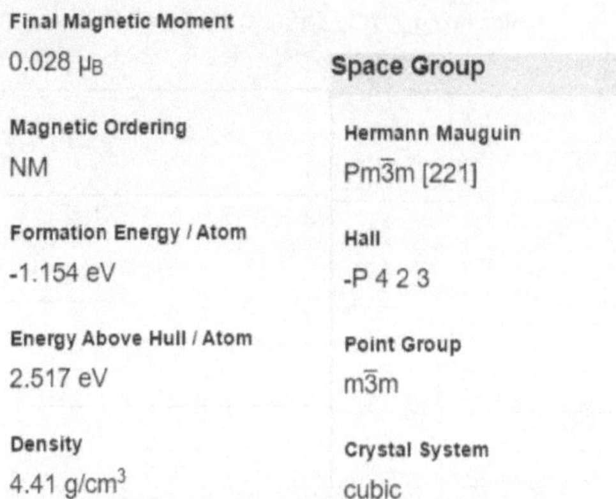

Figure 2. 6. Material Details of $SrZr0_3$

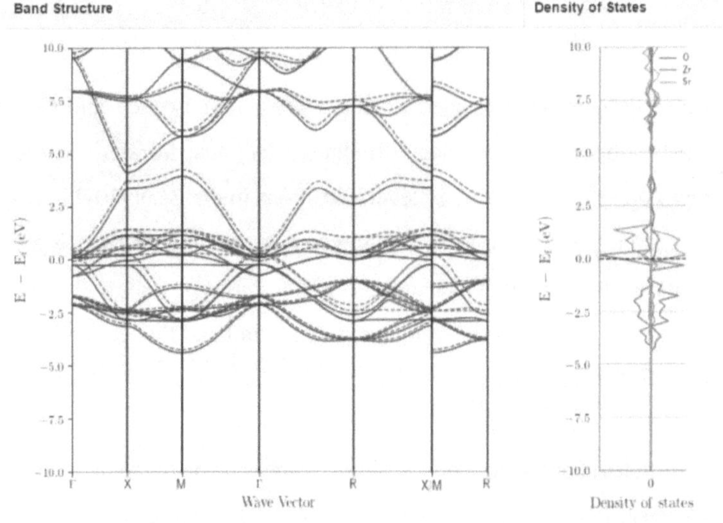

Figure 2. 7. Electronic Structure of $SrZr0_3$

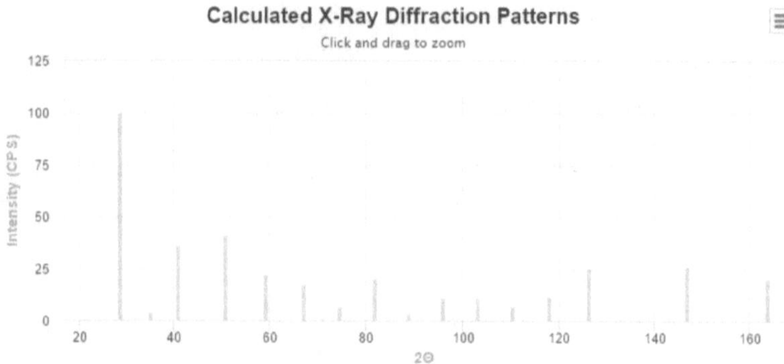

Figure 2. 8. X-ray diffraction patterns of $SrZrO_3$

2.6 PRESENT STATUS OF RESEARCH

Many research groups have studied upconversion luminescence and WL from multiple REIs doped nanopowders irradiated by near-infrared lasers. The luminescence comes from the energy level transitions in the REIs [10-11, 26-29] in a low laser power. Nevertheless, as the power increases higher (above a certain threshold), the excited nanopowder emits a bright white or yellowish light, also called WL, and corresponds to a strong wide emission band that covers the whole visible region [3-4, 10, 55-58].

Hundreds of articles about upconversion luminescence can be found in the literature; however, there are not many papers about WL. Strek et al. reported that they found WL emission in LiP_4O_{12} with Nd^{3+} [59] and Yb^{3+} [57], and AlO_3 with Nd^{3+} [58]. J. Wang et al. reported they found WL emission by using only lanthanide oxides (Sm_2O_3, Tb_2O_3, CeO_2 [55], $Yb_3Al_5O_{12}$ and $(Yb,Y)_2O_3$ [56]). G. Bilir et al. found WL in Nd-doped and un-doped Y_2O_3 nano powders [60], G. Bilir et al. also found

WL in $Y_3Al_5O_{12}$ (YAG) and $Gd_3Ga_5O_{12}$ (GGG) nano powders [63]. And F. González et al. found WL in $Y_4Zr_3O_{12}$ nanopowders with Yb^{3+} [10]. Up to now, there are no reports about $SrZrO_3$ with Er^{3+} or Yb^{3+}, so it gives us a good opportunity to systematically study this material.

2.7 MOTIVATION OF THIS THESIS

This paper is motivated by two reasons. First, the generation for upconversion WL on irradiated nano-crystallines doped with REIs has been discovered for over a decade. The phenomenon was not widely tested on various nano powders and not applied to optics or business industry. WL is different from all the commercial lamps in the current market and has great advantage in nature because the generation of WL through blackbody radiation simulates the sunlight emission process. Commercially, WL is better for eye protection, provides higher lux with lower energy, comparing to other lamps. WL could also be widely used on agriculture by providing stronger artificial photosynthesis. However, the complicated operation process and strict environment condition set great obstacle to commercialization. This study is designed to explore various material samples and contribute another possibility to the literature of WL.

Second, many theoretical questions related to WL generation remain unsolved, such as the principle of generation, the effect of the crystallite size and the role played by some dopant ions on its generation, and if the phenomenon is intrinsic or not to nanostructured systems. This study is also designed to compare with the existing literature on the aspects of environment conditions and sample materials, so that we can further explore the theoretical support to WL generation.

3. PREPARATION OF SAMPLES

The sample was made by *Professor Federico Garcia Gonzalez* in *Universidad Autónoma Metropolitana.*

Polycrystalline ceramic powders were synthesized by the polymerizable complex method, according to the general formula $SrZr_{0.99-x}Er_{0.01}Yb_xO_{2.995-0.5x}$, where the charge balance is attained by oxygen vacancies generation, with x = 0.01, 0.02, 0.04, 0.08 (hereinafter denoted as SZErYb1, SZErYb2, SZErYb4 and SZErYb8, respectively). The starting raw materials used were $Sr(NO_3)_2$ (99%, Sigma-Aldrich), $Zr(O(CH_2)_2\text{-}CH_3)_4$ (70% in 1-propanol, Sigma-Aldrich), $Er(NO_3)_3{\cdot}5H_2O$ (99.9%, Sigma-Aldrich), $Yb(NO_3)_3{\cdot}5H_2O$ (99.9%, Sigma-Aldrich), citric acid $C_3H_4(COOH)_3$ (ACS reagent, Sigma-Aldrich), ethylene glycol $C_2H_6O_2$ (ACS reagent, Sigma-Aldrich) and absolute ethanol (ACS reagent, JT Baker). In a typical synthesis, the zirconium propoxide was dissolved in absolute ethanol under continuous stirring, immediately after that, citric acid (molar ration 1:4, amount of metals: citric acid) and ethylene glycol was added for stabilization of the Zr^{4+} cation. This solution was stirred by 1 h at 60°C. Apart, strontium nitrate and the lanthanide nitrates were dissolved in deionized water under stirring by 30 min at room temperature. This last aqueous solution was mixed with the former alcoholic solution. Then, the temperature of the final solution was raised to 70 °C for induce the evaporation of the solvents resulting in a viscous yellowish solution. For promoting polymerization of the mix, the temperature was set at 90 °C until a resin was obtained. This resin was then pre-calcinated in an oven at 300°C for 1 h obtaining a dark-brown dry foam. To pyrolyze all the organics present, the pre-calcined foam was ground and then powder was annealed at 800°C for 1 h in air, and was further annealed at 900°C and 1100°C for another 1 h in air. Finally, the powders were uniaxially pressed at 100 MPa into disks of 13 mm in diameter and 1 mm in thickness. The pellet was further annealed at 1200°C for 2 h.

The X-ray diffractograms (XRD) of the samples were measured in air and at room temperature using a Bruker D2 Phaser diffractometer working in the Bragg-Brentano θ-θ geometry, with Cu Kα radiation. The apparatus is fitted with a Ni 0.5% Cu−Kβ filter in the secondary beam, and a one-dimensional position-sensitive silicon strip detector (Bruker, Lynxeye). The diffraction intensity as a function of 2θ angle was measured between 15° and 110°, with a 2θ step of 0.014207°, for 38.4 s per point. Crystalline structures were refined using the Rietveld method by using the fundamental parameters approach during the refinements, as implemented in the TOPAS Academic code, version 6. The parameters used in the refinements included polynomial terms for modeling of the background, the lattice parameters, terms indicating the position and intensity of the "tube tails", specimen displacement, structural parameters, and the width of a Lorentzian profile for modeling the average crystallite size. This last feature was modeled in reciprocal space with a symmetrized harmonics expansion. The standard deviations, given in parentheses in the text, show the variation in the last digit of a number; when they correspond to Rietveld refined parameters. Scanning Electron Microscopy (SEM) was used to reveal the morphology of sintered pellets in a JEOL electron microscope. These images were acquired form fresh-fractured specimens.

The XRD patterns of the SZErYb1, 2, 4 and 8 pellets are shown in Figure 3.1. All patterns show the orthorhombic perovskite-like structure corresponding to the undoped $SrZrO_3$ (PDF card 04-014-8276). Also, a small amount ($<$ 5 wt%) of tetragonal ZrO_2 is present for samples SZErYb1, 2 and 4.

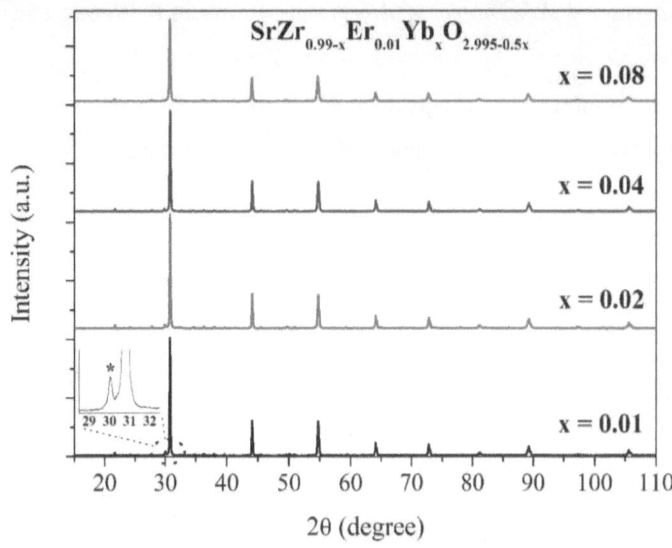

Figure 3. 1. X-ray diffraction patterns of SrZr0.99-xEr0.01YbxO2.995-0.5x pellets.

The inset shows the peak (marked with an asterisk) associated with the (101) plane of the tetragonal ZrO_2 impurity.

To get a deeper understanding about the structural effect for the incorporation of Er^{3+} and Yb^{3+} into the $SrZrO_3$, XRD patterns were analyzed by the Rietveld refinement method. The unit cell of the $Sr\,Zr_{0.99-x}Er_{0.01}Yb_xO_{2.995-0.5x}$ series compound was modelled with an orthorhombic symmetry described by the space group Pbnm, a non-conventional setting of space group 62 but afterward transformed into its Pnma conventional one. According to the stoichiometry, the atomic basis contains $(0.99\text{-}x)Zr^{4+}+0.01\,Er^{3+}+x\,Yb^{3+}$, one Sr^{2+} and two O^{2-} at the relative coordinates (0, 0, 0), (x, y, ¼), (x_{o1}, y_{o1}, ¼) and (x_{o2}, y_{o2}, z_{o2}), respectively.

The initial values for cell parameters and those refinable atomic coordinates were set as those reported in *[A. Ahtee, M. Athee, A.M. Glazer, A.W. Hewat, The structure of orthorhombic $SrZrO_3$ by neutron powder diffraction, Acta Cryst. B32 (1976) 3243-3246]*. To show the good agreement between the experimental X-ray data and the calculated ones, Rietveld refinement plot of the sample SZErYb02 is depicted in Fig. 3.2.

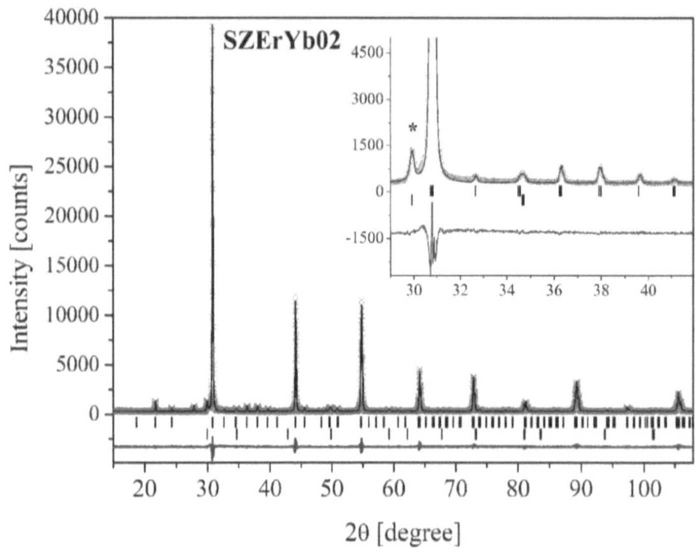

Figure 3. 2. Rietveld refinement plot of the pellet SZErYb02.

The scatter points (red) and the upper solid line (black) correspond to the experimental and calculated data, respectively. The lower curve (blue) is the difference between calculated and measured diffraction patterns. Vertical marks on the bottom, represent the Bragg reflections associated with the rhombohedral perovskite-like (upper) and the tetragonal ZrO_2 (lower) phases. In the inset is shown the two-theta range from 29 ° to 42 °, where again, the peak associated with the (101) plane of the tetragonal ZrO_2 is indicated by an asterisk.

The results of the Rietveld analyses are presented in the Table 3.1; they include the lattice parameters and the average crystallite size, as a function of the Yb^{3+} content (x). Also, in Table 1 it is shown the concentration in wt % of the tetragonal ZrO_2 crystal, and the figures of merit, Rwp, for the refinements, which further support the excellent agreement between experimental and calculated data, as above-mentioned.

Table 3.1. Rietveld refinement results for the $SrZr_{0.99-x}Er_{0.01}Yb_xO_{2.995-0.5x}$ pellets

x	Lattice parameters (Å)[a]			Cryst. size	ZrO_2 content	R_{wp}
(mol)	a (Å)	b (Å)	c (Å)	(nm)	(wt%)	(%)
0.01	5.81668(7)	8.20546(10)	5.79656(7)	216(2)	4.6(7)	6.78
0.02	5.81739(8)	8.20611(10)	5.79666(7)	188(1)	2.9(8)	6.71
0.04	5.81767(9)	8.20711(12)	5.79722(8)	145(1)	2.9(8)	6.44
0.08	5.82340(13)	8.21147(20)	5.80257(13)	88(1)	-	6.90

Interesting to notice from data in Table 3.1, are the changes in the lattice parameters and average crystallite size as the Yb^{3+} content increases. The enlargement of the lattice parameters is consistent with the substitution of the Zr^{4+} at the VI-coordinated B site of the perovskite-like structure by the larger Er^{3+} and Yb^{3+} ions. According to the well-known ionic radii values reported by Shannon *[R. D. Shannon, Revised effective ionic radii and systematic studies of interatomic distances in halides and chalcogenides, Acta Cryst. A32 (1976) 751-767]*, in coordination VI the Zr^{4+}, Er^{3+} and Yb^{3+}, possess radii of 0.72 Å, 0.89 Å, and 0.87 Å, respectively. On the other hand, the decrease in crystallite size is related to the distortions generated in the crystalline structure by the incorporation of Er^{3+}/Yb^{3+} into the $SrZrO_3$ lattice. These distortions which include oxygen vacancies, increase as the amount of Yb^{3+} increases, and in some way, impose limitations to the growth of crystallites.

In Fig. 3.3, SEM micrographs for all samples are depicted. It can be seen particles uniformly distributed and crystallites having uniform shapes and sizes. A more detailed observation of the images, allows to realize that the crystallite size decreases as the Yb3+ concentration increases accordingly to the Rietveld refinement analyses.

Figure 3. 3. SEM images obtained from fractured pellets. SZErYb01 (a), SZErYb02 (b), SZErYb04 (c), and SZErYb08 (d).

4. EXPERIMENTAL APPARATUS

4.1 LIGHT SOURCE

a). Diode Laser

Figure 4.1 shows the Laser Drive Inc. HF-975-010 diode laser, made by Axel Photonics. It is used for the continuous wave (CW) excitation of the samples at 975 nm wavelength, and generated by a function generator which can produce a maximum current of 10 A. The output power range of the laser is from 0.3 W to 6.3 W.

Figure 4. 1. Laser Drive Inc. HF-975-010 diode laser

b). Argon Laser

Figure 4.2 shows the Argon-ton laser system, the Omnichrome model 532, made by Melles Griot company. This CW air-cooled laser is capable of providing 300 mW of output power, low noise, and long plasma tube life. The instrument is a passive column gas discharge laser using smugly ionized argon as the optical gam medium. Its plasma tube is a rugged metal-ceramic device. The resonator is a cast aluminum alloy to stable performance and pointing stability. Laser output is produced in nine wavelengths, from 454 to 514 nm, and can be selected. In our experiments all wavelengths are used.

Figure 4. 2. Argon-ton laser system, Omnichrome model 532

c). White Light Laser

Figure 4.3 shows the HL-3 plus-CAL Calibrated Light Sources for the VIS-Shortwave NIR (350 nm-1700 nm). It is a tungsten-halogen light source that provides known absolute intensity values at several wavelengths. Since the spectral intensity of the HL-3 Series can be traced to an intensity standard traceable to the National Institute of Standards and Technology (NIST), it is specifically designed for calibrating the absolute spectral response of the detecting system. Figure 4.4 shows the output spectrum of HL-3-CAL Calibrate light. The light source was used to create the sensitivity curve of the photomultiplier for the spectrum correction in our experiment.

Figure 4. 3. HL-3 plus-CAL Calibrated Light Sources

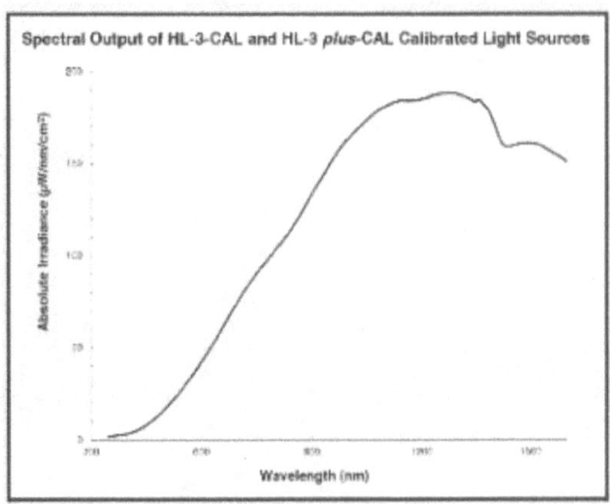

Figure 4. 4. Output spectrum of HL-3-CAL Calibrate light

The corrected curve is created by the following equation:

$$C \times S = U$$

$$C = \frac{U}{S}$$

Where C is the corrected spectra, U is the uncorrected spectra, and S is the sensitivity curve.

4.2 DETECTORS

a). Monochromator

Figure 4.5 shows the McPherson model 2051 one-meter scanning monochromator. The instrument is fitted with a 600 G grating groove, blazed at 1.25 μm. It has a resolution and wavelength resettability of 0.1 A with repeatability of +/- 0 05 A. The dispersion is 8.33 A/mm with a 1 200 G/min grating. The scan controller used is model 789A-3, allowing to set the scanning speed, manually with Thumbwheel, from 0.1 to 999.9 angstroms per minute. Samples are scanned to a maximum wavelength of 15740 A. The entrance and exit slits are adjustable from 5 microns to 2 mm.

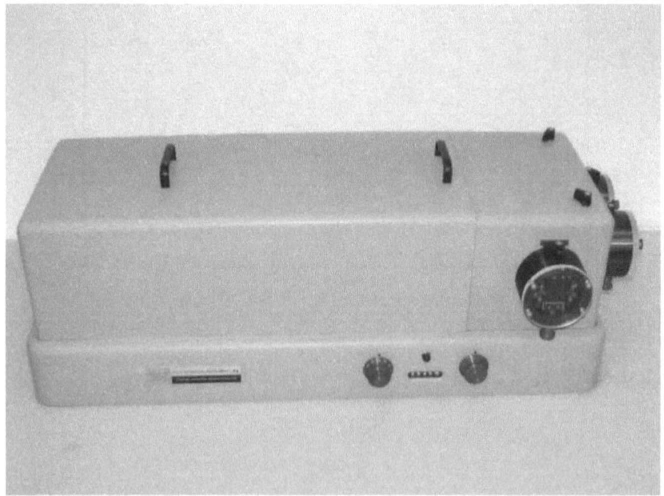

Figure 4. 5. McPherson model 2051 one-meter scanning monochromator

b). Infrared Detector.

The Kolmar Technologies Model KJSDP-1-Jl/DC Infrared (IR) detector is used in continuous optical signal is the Indium- Antimonite (lnSb) type. This is a sensitive and fast (7 ns) photodiode with useful spectral range from 700 μm to 5.4 μm. The detector is a 1x1 mm photodiode integrated with a pre-amplifier whose bandwidth is from DC to 15 MHz. The responsivity of the detector is better than 2x 10E4 V/W and a spectral response (D*) of l EJ.

The Dewar can be funnel-filled with liquid nitrogen that can hold the temperature for 12 hours.

c). Photomultiplier Tube (PMT).

Figure 4.6 shows the photomultiplier tube. The types that we used are the Hamamatsu R1387 and 7102. The useful spectral response is from 300 to 850 nm (S-20) curve for the Rl387, and 400 to 1200 nm (S-1) for the 7102. The PMT is powered by a bench-top variable power supply. The voltage is adjusted, as required, to prevent saturation of the output. The output current of the PMT is proportional to the voltage applied to the bleeder ladder network. Cooling for the PMT is provided by a thermoelectric cooler that cools the tube to about 50 deg. C below ambient.

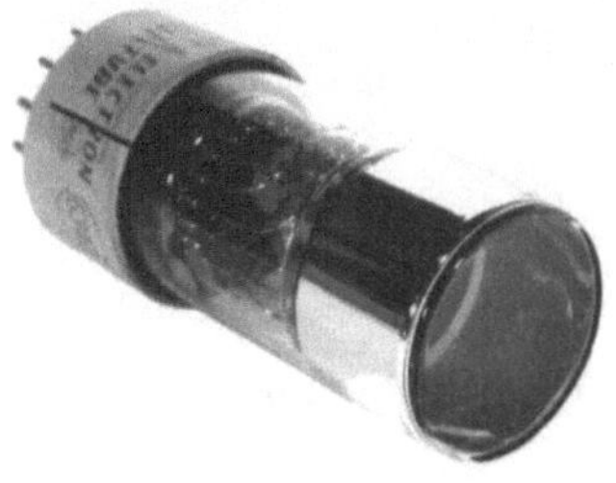

Figure 4. 6. photomultiplier tube Hamamatsu type R1387

d). oscilloscope

Figure 4.7 shows the Rigol DS1054Z Digital Oscilloscope. It has 50 MHz DSO 4 Channels and 50 MHz analog bandwidth total of 4 analog channels. Its maximum waveform capture rate of up to 30,000 wfms/s. The DS1104Z-S Plus also provides the ability to quickly focus in on certain piece of a recorded waveform. It is used to measure the rise and decay pattern of light emission.

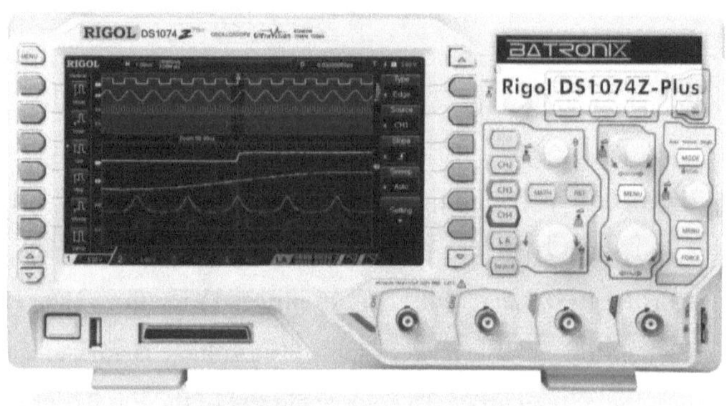

Figure 4. 7. Rigol DS1054Z Digital Oscilloscope

e). Illuminance Meter ASP-MK350

Figure 4.8 shows the Pro ASP-MK350 Model Illuminance Meter (a spectrophotometer) made by Allied Scientific PRO is calibrated to measure illuminance and spectral response of all kinds of light sources. Compare with the general spectrophotometer, it is safe, quick and convenient. Many parameters such as wavelengths, intensity and color of LED can be provided directly through the meter. Besides, it can provide other optical parameters like chromaticity coordinates CIE1931(x, y), CIE1976(u', v'), CCT (correlated color temperature), CRI (color rendering index), light intensity (illuminance), wavelength of peak, and peak counts.

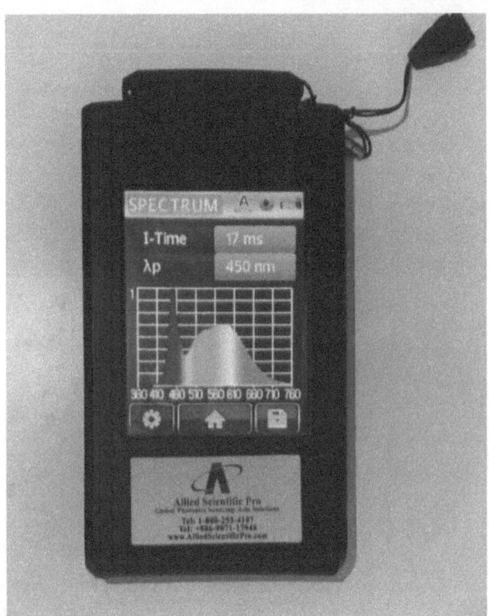

Figure 4. 8. Pro ASP-MK350 Model Illuminance Meter

4.3 SAMPLE ENVIRONMENT

a). Cryogenic Cooler

Figure 4.9 shows the cryogenic refrigerator, which operates on the Gifford-McMahon (GM) principle using a closed helium gas cycle. The advantage of the

GM is that the compressor unit can be separated from the Cold Head which is part of the Sample Chamber, thus allowing the flexibility of mounting the Cold Head in any position. The compressor and the Cold Head are connected with pressure flexible tubing. The system is filled with helium to a pressure of 16 bar, capable of cooling the sample to 20 K.

Figure 4. 9. helium gas cyclic cryogenic refrigerator

b). Vacuum System

The Vacuum System lowers the pressure of the Sample Chamber to about 2x 10 s Torr in two stages. In the first stage a mechanical 'roughing' pump achieves about 30 microns of vacuum. The pressure is further lowered by a second stage consisting of a compact air-cooled diffusion pump, capable of Ix 10 6 Torr for small volumes.

c). Vacuum Pump

Figure 4.10 shows the T-Station fully automatic pumping system, which is suitable for a wide range of vacuum applications. The T-Station can be supplied with either an XDD1 oil free diaphragm pump or an E2M1.5 rotary vane pump, both system variants use an EXT75DX turbomolecular pump. The T-Station is controlled by an easy to use touch pad control module. A single gauge input included can be connected to a range of Edwards active gauges allowing for pressure measurement and or control management of the turbomolecular pump. The compact size of the T-Station is ideal for use on bench-tops or suitable mobile platforms. The open system configuration allows easy maintenance of the main pump components.

Figure 4. 10. T-Station vacuum pumping system

d). Sample Chamber.

The sample chamber was manufactured by Janis. The sample can be easily mounted on a pedestal and adjusted for orientation with the Monochromator input FOV and the light source beam. It has five optical windows for versatility of beam positioning and steering. The sample can be cooled to about 20 K in a vacuum of 10^{-5} Torr. Because of the vibration and noise generated by the cryogenic pump, the sample chamber is mounted on a concrete column standing on the floor, weighing over 200 pounds. The column is adjustable for the correct sample position in the line of sight with the monochromator input slit.

e). temperature controller

Figure 4.11 shows the Lake Shore Cryotronics 331 Model temperature controller, which controls the ambient temperature at a range of 50K-800K.

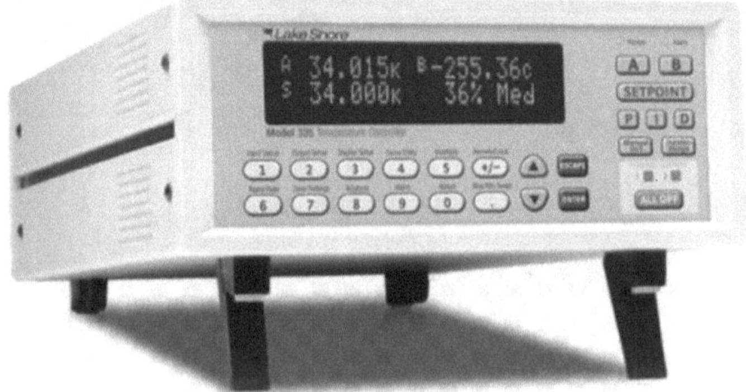

Figure 4. 11. Lake Shore Cryotronics 331 Model temperature controller

4.4 SIGNAL CONDITIONING

a). Pre-Amplifier

The output current of the PMT anode is read, as a voltage, with a load resistor with respect to ground potential. The effective bandwidth (BW) of the output pulse is inversely proportional to the product of the resistor and all parasitic capacitance in the PMT output circuit, including cabling. Thus to increase the BW an amplifier is put close to the PMT output and the load resistor is made as small as possible. The pre-amplifier is home-made using an ultra-low distortion, high speed integrated circuit, the Analog Devices AD 8008. The chip has a BW specification of 230 MHz for a voltage gain of 2. With the pre-amplifier in place the PMT output is 50 to 1000 ohms, depending on the output signal strength. The pre-amp gain can be two or 20. The pre-amp is powered by AA cells.

b). Chopper

The chopper is placed at the entrance of the monochromator continuous wave (CWJ optical signal in essence the mechanical chopper, operated at nominally 250 Hz, becomes earner frequency of the optical signal. thus removing all the DC biases introduced by the instrumentation and amplifiers.

c). Lock-in Amplifier

The purpose of this instrument is to amplify the low level AC signal of the PMT output, when the signal is chopped, and demodulate it to recover the base-band optical signal. In this set-up we use the EG&G Model 5201, which is shown in Figure 4.12. It has a 1 μV to 5V rms input sensitivity, with a carrier frequency of 0.2 Hz to 200 kHz. Signal to reference phase shift control can be in 0.025 degree steps from 0 to 360-degree C.

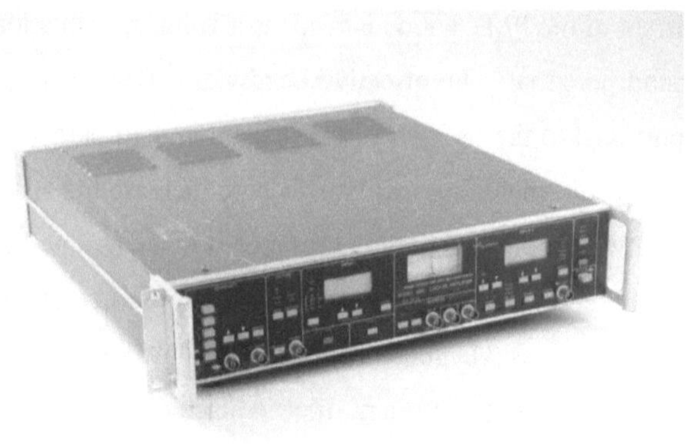

Figure 4. 12. EG&G Model 5201 Lock-in Amplifier

4.5 DATA ACQUISITION

a. Box-Car Sampling System.

While there are new computer methods of acquiring continuous pulsed signals, the Spectroscopy Laboratory is equipped with a dependable and proven signal sampling system known as Box-Car Averager. It is basically a sample-and-hold circuit which integrates the signal during a sampling window and stores it. The sampling window and its position in the signal can be selected as desired. For repetitive signals the sampling window is shifted forward at a new portion of the signal, during the next cycle. A set of samples is thus accumulated during the time of interest in the signal. The sampling window widthand the delay is set on the panel of the Box-Car Averager. In this manner background noise is not integrated because the signal is captured only during the time of interest, and the sample-and-hold integrator removes fluctuations, or noise, during this time. This method of signal capturing is particularly suitable when the repetitive signal is a very small fraction of the duty cycle. The signal is captured with a set of samples only m the mterv al of interest, and then averaged over many cycles.

The Instruments used are the Signal Recovery Model 41218. with the dual AID converter module 4161A This system is capable of 1.5 ns sampling gate width. and an input BW of 450 MHz. and a signal repetition rate of 80 kl+z, maximum. The Spectroscopy Laboratory has recently acquired Data Acquisition Software by National Instrument, called LabVIEW. The instruments above can be interfaced with LabVIEW compatible software dnvers so that the data logging will be more automated. The gate width and delay can also be controlled by Lab VIEW software with Delay Generator Model 9650A.

b. Data Logging Computer.

The output of the Box-Car Averaging System is sent to a Data Logging computer that is programmed to output plots of intensity vs wavelength. These plots are formatted to be published in reports. When we shaJI introduce Lab VIEW software the whole data Jogging control and programming will be done using a graphical interface to the program.

5. LASER POWER AND Yb^{3+} CONCENTRATION EFFECTS ON THE UPCONVERSION LUMINESCENCE EMISSION BY Er^{3+}-Yb^{3+} DOPED $SrZrO_3$ NANO CRYSTALLINE

5.1 INTRODUCTION

In recent years, the upconversion luminescence in many phosphors doped with trivalent rare earth ions (REI) has become an important topic in the luminescence research area. Due to the immense potential applications in materials [61], solar cells [62], and other chemical and biological study, many researchers devote themselves to the exploration of the effects and properties of all kinds of hosts as well as their dopants. There are many good hosts like Y_2O_3 [19], LaF_3 [54], Gd_2O_3 [11] and so on.

Rare-earth oxides (REO) [2-6] phosphors have been widely studied because of their good thermal properties and physical/chemical stability [20-21]. Based on the 975 nm diode laser (which has $\sim 10\times 10^3$ cm^{-1} photon energy) in our lab, the most efficient upconversion phosphors are the Er^{3+}/Yb^{3+} co-doped REO nano crystalline due to the same transition energy ($\sim 10\times 10^3$ cm^{-1}) in the $^4I_{15/2} \rightarrow {}^4I_{11/2}$ levels of Er^{3+} and the $^2F_{7/2} \rightarrow {}^2F_{5/2}$ levels of Yb^{3+} ions. There are several publications about REO upconversion luminescence in the literature review such as Er^{3+}-Yb^{3+} co-doped Y_2O_3 [19] and Er^{3+}-Yb^{3+} co-doped Gd_2O_3 [11], however, until now there is no report about the $SrZrO_3$ system, and this gives us a good opportunity to work on it.

The luminescence phenomenon is generally affected by a change of temperature [11], so studying the variation of the temperature in nano crystalline can help us to understand the mechanisms responsible for certain characteristics luminescence, such as its quantum efficiency and its potential for applications. In our work, the temperature dependence on the laser power and Yb^{3+} concentration was investigated. With the excitation from a 975 nm diode laser, the emission spectra (shown in Figure 5.1 in section 5.3) of the sample: $SrZr_{0.99-x}Er_{0.01}Yb_xO_{2.995-0.5x}$ ($x = 0.01, 0.02, 0.04, 0.08$) were obtained. A clear green and red emission due to the energy transitions of the excited states $^2H_{11/2}$, $^4S_{3/2}$ and $^4F_{9/2}$ to the ground state $^4I_{15/2}$ of Er^{3+} ions were observed, as well as the characteristic maxima of the spectra.

According to the statistical mechanics, the occupation probability of an excited-electronic population depends on the temperature through the Boltzmann factor: e^{-E/k_BT} [11, 54], where E is the energy of each state, k_B is the Boltzmann constant, and T is the absolute temperature. Considering the Er^{3+} emission in our measurement, the populations of the two excited states $^2H_{11/2}$ and $^4S_{3/2}$ are linked through the expression $N_{^2H_{\frac{11}{2}}} = N_{^4S_{\frac{3}{2}}}(g_{^2H_{\frac{11}{2}}}/g_{^4S_{\frac{3}{2}}})e^{-\Delta E/k_BT}$, where $N_{^2H_{11/2}}$ and $N_{^4S_{3/2}}$ stands for the electronic population of the levels $^2H_{11/2}$ and $^4S_{3/2}$, and g stands for the degeneracy of each state. Moreover, the population of each state also corresponds to the emission intensity in the luminescence process, given by $I_{^{2S+1}L_J} = f_{^{2S+1}L_J} N_{^{2S+1}L_J}$, where $I_{^{2S+1}L_J}$ corresponds to the emission intensity of the level $^{2S+1}L_J$, and $f_{^{2S+1}L_J}$ is a constant factor. For the energy transition from the two excited states $^2H_{11/2}$ and $^4S_{3/2}$ to the ground state $^4I_{15/2}$, the integrated emission intensity ratio

$I_{^2H_{11/2}}/I_{^4S_{3/2}}$ can be easily calculated from the spectra, and this ratio is called luminescence intensity ratio (LIR). Thus, we provide the way to relate the temperature with the spectra through the equation:

$$LIR=\frac{I_{^2H_{11/2}}}{I_{^4S_{3/2}}}=\frac{f_{^2H_{11/2}}N_{^2H_{11/2}}}{f_{^4S_{3/2}}N_{^4S_{3/2}}}=\frac{f_{^2H_{11/2}}}{f_{^4S_{3/2}}}e^{-\Delta E/k_BT}=Be^{-\Delta E/k_BT} \quad (1)$$

where ΔE is the energy difference between the two levels (Figure 2.4), 750 cm^{-1}, k_B is the Boltzmann constant, T is the absolute temperature, and B is a constant.

After determining the temperature of the emitting sample through the experimental results, namely, LIR from the spectra, we obtained the information on the sample temperature as a function of the laser power and Yb^{3+} concentration in the nano crystalline. This also gives us a better understanding of White Light (WL), because our results also show the appearance of continuous WL emission under high power laser irradiation and Yb^{3+} concentration, while luminescence comes through the low power. Some other experimental evidence [58-60] also seems to indicate that the WL emission is thermal in origin, and we wish to further examine this possibility.

5.2 EXPERIMENT

The sample $SrZrO_3$ nano pellets co-doped with 1% Er^{3+} and 1, 2, 4, 8% Yb^{3+} (SZErYb1, 2, 4, 8) were synthesized by *Professor Federico Garcia Gonzalez* in Universidad Autónoma Metropolitana in Mexico.

The upconversion emission spectra were measured at room temperature (300K) and vacuum condition with a pressure of 0.001 mbar. The sample was pumped by a laser diode power supply LDI-820 with emitting wavelength at 975 nm and maximum output of 6.79 W, and the luminescence signal was directed to the entrance slit of an Ocean Optics USB4000 spectrometer which was fitted with a 3648-element linear silicon CCD array detector, and recorded in the range from 500 nm to 700 nm by a computer. A 900 nm short-pass filter was used in front of the detector to block the signal from the laser.

For comparison, the sample was also pumped by an air-cooled Argon-ton Laser (454 to 514 nm), Omnichrome model 532, with nine emitting wavelengths (454 to 514 nm) and 300 mW of output power. A 475 short-pass filter was used in front of the laser, and a 500 long-pass filter was used in front of the detector to block the signals from the laser. A closed cycle Helium refrigerator was used to produce the environment temperature at a range of 50-800 K, and the temperature was controlled by a Lake Shore Cryotrons 331 Model temperature controller.

5.3 RESULTS AND DISCUSSION

Figure 5.1 shows the upconversion emission spectra of the pellets SZErYb1, 2, 4, 8 under 975 nm laser irradiation with different laser powers at a range of 1.54 W to 6.28 W. The spectra of the sample SZErYb1, 2, 4 (see Figure 5.1 (a), (b), (c)), show the luminescence emission maxima at 520 nm, 550nm and 660 nm, namely, green and red light, associated with the Er^{3+} energy level transitions of excited states $^2H_{11/2}$/ $^4S_{3/2}$ and $^4F_{9/2}$ to the ground state $^4I_{15/2}$, respectively. The spectrum of the sample SZErYb8 (see Figure 5.1 (d)), shows luminescence emission at a lower power (1.54 - 4.81 W), and WL emission at a higher power (above 5.55 W); the luminescence intensity (both green and red emission) increased to a maximum value at 3.21 W, and then decreased as the power increased, until WL is seen (5.55 W). The decrease in the luminescence intensity was likely produced by the non-radiative processes activated by the higher temperature in the sample. More details about WL emission will be discussed in Chapter 6. Moreover, the sample SZErYb2 exhibits the highest emission intensity (Figure 5.1(b)).

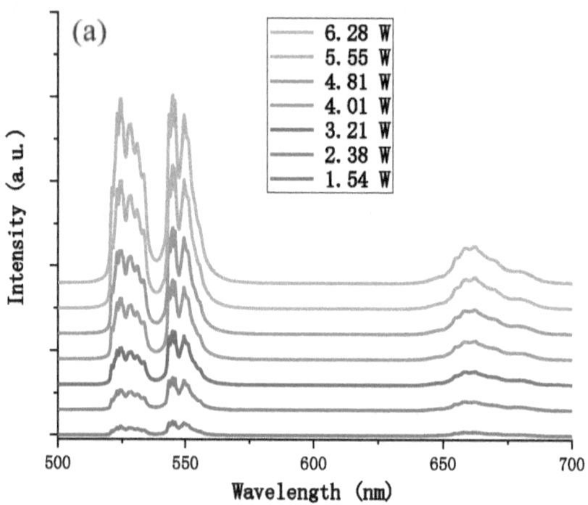

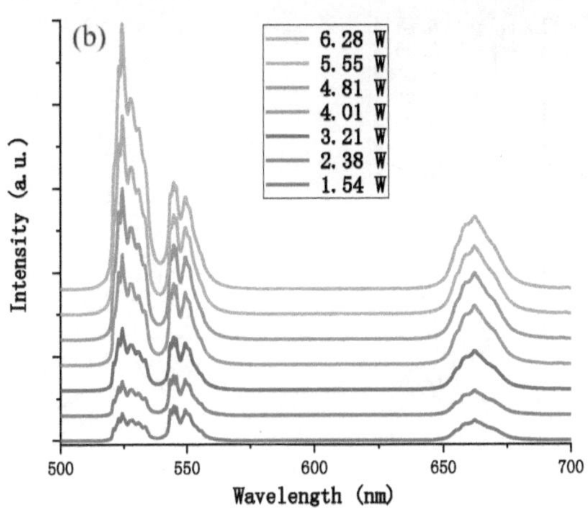
(b)
6.28 W
5.55 W
4.81 W
4.01 W
3.21 W
2.38 W
1.54 W
Intensity (a.u.)
500
550
600
650
700
Wavelength (nm)

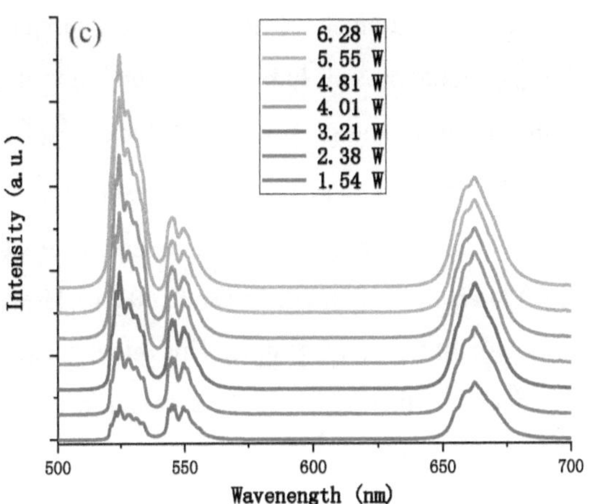
(c)
6.28 W
5.55 W
4.81 W
4.01 W
3.21 W
2.38 W
1.54 W
Intensity (a.u.)
500
550
600
650
700
Wavenength (nm)

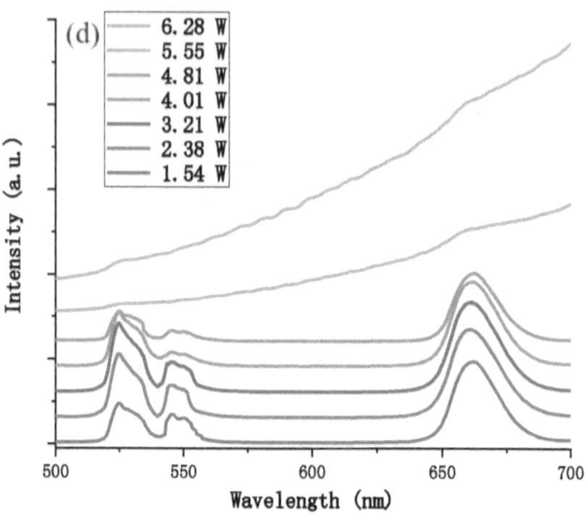

Figure 5. 1. Upconversion luminescence emission spectra of pellets SZErYb1 (a), SZErYb2 (b), SZErYb4 (c) and SZErYb8 (d), measured at different excitation laser powers.

We also tried the same measurement of the $SrZrErO_3$ sample without Yb^{3+}. Theoretically speaking, it should provide luminescence emission at the visible range, because the electronic transition energy from excited state $^4I_{15/2}$ to ground state $^4I_{11/2}$ of Er^{3+} is 10×10^{-3} cm^{-1} is very close to the excitation photon energy by the 975 nm infrared laser, and this may also cause the upconversion luminescence from infrared (975 nm) to visible light (520, 550 and 660 nm). However, at this time we didn't see any visible light from this sample. This might because the absorption efficiency of Er^{3+} is much less then Yb^{3+}, so the subsequent upconversion process can hardly happen.

In order to further characterize the luminescence features of the spectra shown in Figure 5.1, we plotted the normalized integrated intensities (Figure 5.2). The LIR variation (Figure 5.3) of the three emission bands as a function of the laser power, respectively, for the four different Yb^{3+} concentration. The data plotted in Figures 5.2 and 5.3 are given in Table 5.1 – 5.4.

Power(W)	1.54	2.38	3.21	4.01	4.81	5.55	6.28
$^2H_{11/2}$	0.023	0.072	0.125	0.209	0.301	0.441	0.623
$^4S_{3/2}$	0.041	0.109	0.161	0.220	0.318	0.433	0.579
$^4F_{9/2}$	0.033	0.059	0.101	0.125	0.170	0.201	0.248
LIR	0.561	0.661	0.776	0.950	0.947	1.018	1.076

Table 5. 1 Normalized integrated intensities of the energy level transitions of Er^{3+} as a function of the laser power for sample SzErYb1.

Power(W)	1.54	2.38	3.21	4.01	4.81	5.55	6.28
$^2H_{11/2}$	0.055	0.108	0.233	0.329	0.475	0.625	0.849
$^4S_{3/2}$	0.081	0.141	0.213	0.265	0.317	0.349	0.396
$^4F_{9/2}$	0.083	0.139	0.234	0.301	0.348	0.362	0.401
LIR	0.679	0.765	1.093	1.241	1.498	1.790	2.143

Table 5. 2 Normalized integrated intensities of the energy level transitions of Er^{3+} as a function of the laser power for sample SzErYb2.

Power(W)	1.54	2.38	3.21	4.01	4.81	5.55	6.28
$^2H_{11/2}$	0.093	0.209	0.323	0.437	0.552	0.679	0.737
$^4S_{3/2}$	0.105	0.183	0.221	0.242	0.247	0.255	0.250
$^4F_{9/2}$	0.270	0.404	0.509	0.579	0.570	0.591	0.585
LIR	0.885	1.142	1.461	1.805	2.234	2.662	2.948

Table 5. 3 Normalized integrated intensities of the energy level transitions of Er^{3+} as a function of the laser power for sample SzErYb4.

Power(W)	1.54	2.38	3.21	4.01	4.81	5.55	6.28
$^2H_{11/2}$	0.059	0.078	0.094	0.063	0.046		
$^4S_{3/2}$	0.042	0.045	0.042	0.020	0.012		
$^4F_{9/2}$	0.299	0.362	0.337	0.305	0.279		
LIR	1.405	1.729	2.210	3.050	3.770		

Table 5. 4 Normalized integrated intensities of the energy level transitions of Er^{3+} as a function of the laser power for sample SzErYb8.

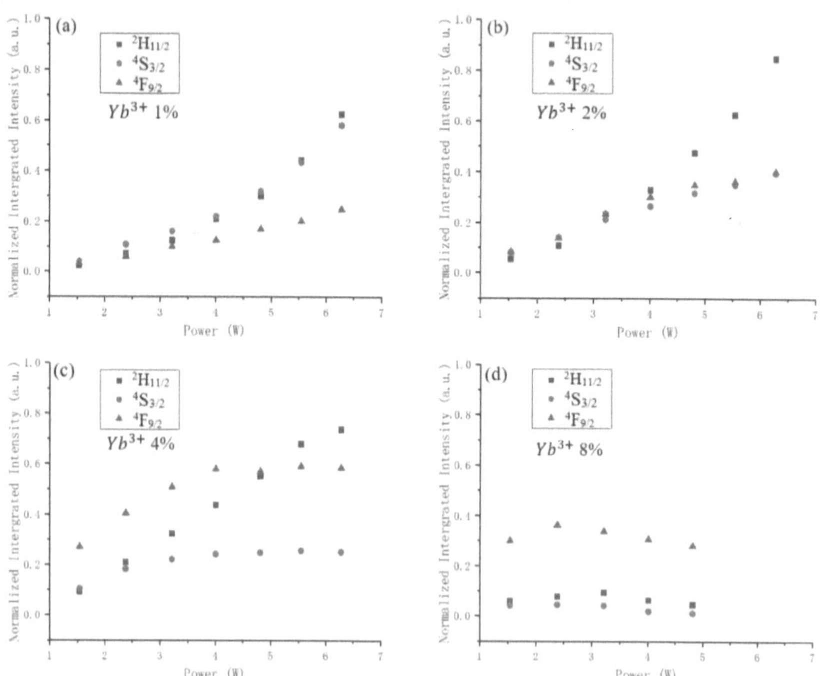

Figure 5. 2. Normalized integrated intensities of the energy level transitions of Er^{3+} ($^2H_{11/2}$, $^4S_{3/2}$ and $^4F_{9/2}$ excited states to $^4I_{15/2}$ ground state) as a function of the laser power for the samples SZErYb1 (a), SzErYb2 (b), SzErYb4 (c) and SzErYb8 (d).

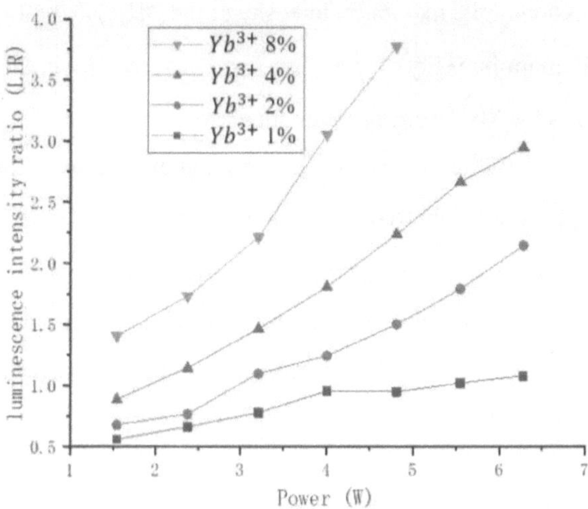

Figure 5. 3. luminescence intensity ratio (LIR) as a function of the laser power for the samples SzErYb1, SzErYb2, SzErYb4 and SzErYb8.

From the figures above, we summarize our experimental findings as follows:

1), For the luminescence, as the laser power increases, the emission intensity of the maxima of the $^2H_{11/2} \rightarrow {}^4I_{15/2}$ transition, always increase, compared with the emission intensity maxima of the $^4S_{3/2} \rightarrow {}^4I_{15/2}$ transition. Thus, the LIR always increases as the laser power increases.

2), At a certain laser power, when the Yb^{3+} concentration increases, the emission intensity of the maxima of the $^2H_{11/2} \rightarrow {}^4I_{15/2}$ transition, always increase, compared with the emission intensity maxima of the $^4S_{3/2} \rightarrow {}^4I_{15/2}$ transition, and the LIR also increases.

3), When the emission of the maximum intensity of the $^2H_{11/2} \rightarrow {}^4I_{15/2}$ transition equals to the maximum intensity of the $^4S_{3/2} \rightarrow {}^4I_{15/2}$ transition, the laser power always decreases as the Yb^{3+} concentration increases.

4), As the Yb^{3+} concentration increases, The emission of the maximum intensity of the $^4F_{9/2} \rightarrow {}^4I_{15/2}$ transition (red light), always increases, compared with the emission of the maximum intensity of the $^2H_{11/2}/\,{}^4S_{3/2} \rightarrow {}^4I_{15/2}$ transitions (green light), and finally becomes dominant for all laser power in sample SZErYb8. (We do see a green color emission from sample SrZrYb1 and 2, orange color from sample SrZrYb4, and red color from sample SrZrYb8 by eyes)

Based on eq. (1), the temperature T changes monotonically with LIR in all four samples, so it also increases as the laser power increases. On the other hand, when the power is fixed, the temperature T always increases as the Yb^{3+} concentration increases, and culminates until the continuous WL is seen in the sample SZErYb8 (Figure 5.1(d)). The decrease of the luminescence in sample SZErYb8 indicated that when more energy from the laser comes into the system, less energy is released as light. Moreover, as the concentration of the absorber entities (Yb^{3+}) increases while the emitter entities (Er^{3+}) stays constant, the light energy decreases as well. These facts suggest that, the non-radiative processes increase as the temperature increases, and the luminescence intensity tends to decrease, after reaching a maximum in the sample SZErYb2, as the Yb^{3+} concentration increases. Finally, continuous WL emission, which is similar to thermal-like radiation, appeared only in a higher temperature condition.

In order to get a quantitative observation for the variance of the temperature related to the laser power and the Yb^{3+} concentration, we used eq. (1) to calculate the temperature's values.

$$LIR=\frac{I_{^2H_{11/2}}}{I_{^4S_{3/2}}}=\frac{f_{^2H_{11/2}}N_{^2H_{11/2}}}{f_{^4S_{3/2}}N_{^4S_{3/2}}}=\frac{f_{^2H_{11/2}}}{f_{^4S_{3/2}}}e^{-\Delta E/k_BT}=Be^{-\Delta E/k_BT} \quad (1)$$

First we need to calculate B in the function above. When the laser power is 0, the temperature is 300 K which stands for the room temperature, the corresponding LIR value can be obtained by extrapolating the linear fitting curve in the LIR ~ power plot (see Figure 5.3) to the limit zero. Thus, we have

$$B = (LIR)_0/e^{-\Delta E/k_BT_0} \quad (2)$$

Where $\Delta E = 750\ cm^{-1}$, k_B is the Boltzmann constant, $T_0 = 300K$, and $(LIR)_0$ is the corresponding LIR value at 0 power.

After getting B, we can easily calculate the temperature of each LIR, i.e. each laser power:

$$T = -\Delta E\ /\ (k_B ln\frac{LIR}{B}) \quad (3)$$

The plots of the calculated temperature T versus laser power are shown in Figure 5.4 below, with the corresponding data given in Table 5.5.

T (K)	Power (W)						
Sample	1.54	2.38	3.21	4.01	4.81	5.55	6.28
SzErYb1	313	330	347	351	362	369	375
SzErYb2	330	349	380	411	435	472	518
SzErYb4	374	415	450	489	548	592	620
SzErYb8	470	557	651	740	785		

Table 5. 5 Calculated temperature (from eq. 3) as a function of the laser power, for samples SzErYb1, 2, 4 and 8.

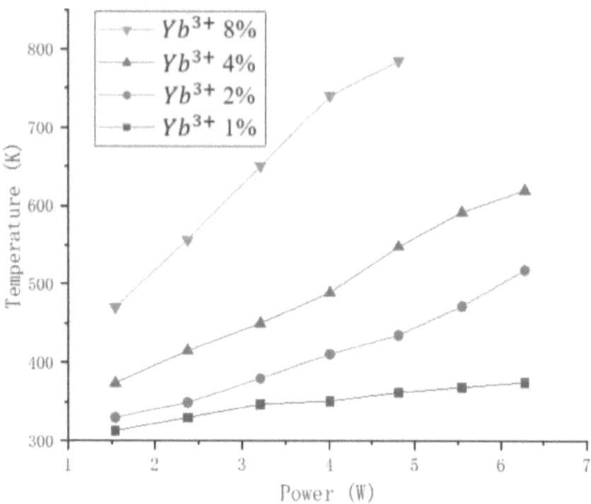

Figure 5. 4. Calculated temperature (from eq. 3) as a function of the laser power, for samples SzErYb1, 2, 4 and 8.

From the Figure 5.4, we can see that

1), The temperature T of the sample increases as the laser power increases, at a certain fixed concentration of Yb^{3+};

2), The temperature T of the sample increases as the concentration of Yb^{3+} increases, at the same laser power.

The presence of Yb^{3+} increases the absorption efficiency of the excitation photon energy from the laser into the samples, and this process favors the rise of temperature, as well as the generation of WL, but not always for the luminescence. On the other hand, the increasing of Yb^{3+} concentration also increases the non-radiative losses, and this process causes a decrease in the luminescence effect. At the same time, however, the increasing of Yb^{3+} increases the absorption of the light energy coming from the laser into the samples as well. Thus, both mechanisms indicate that a high

Yb^{3+} concentration leads to a high rate of heating of the sample, and the temperature of the sample continues increasing until WL appears. All these results show an evidence that the WL is induced thermally.

For purpose of further verification of the relationship between luminescence LIR and the sample temperature, we set up another experiment with sample SZErYb2 irradiated by a nine wavelengths Argon Laser (454 to 514 nm) for comparison. Instead of varying the laser power, we change the environment temperature T_0 at a fixed power. The sample also provides the same luminescence (green and red light) by irradiation of Argon laser (but not upconversion, because the excitation wavelength is less than the emission wavelength). The power of the Argon laser is very low, so it doesn't heat the sample, and the temperature T of the sample is equal to the temperature T_0 in the environment. Figure 5.5 shows the LIR-T plots of samples at irradiation of the Argon Laser. For comparison, Figure 5.6 shows the LIR-T plots for samples SZErYb1, 2, 4 and 8 under the irradiation of the diode laser. Figure 5.7 shows the comparison of the LIR – T plots of sample SZErYb2 under the excitation of both lasers. We see that the variation tendency of LIR from the 975 diode laser shows a good agreement with the Argon laser. Thus, the results from the experiment using the Argon laser show that determining the temperature from the LIR data is reliable.

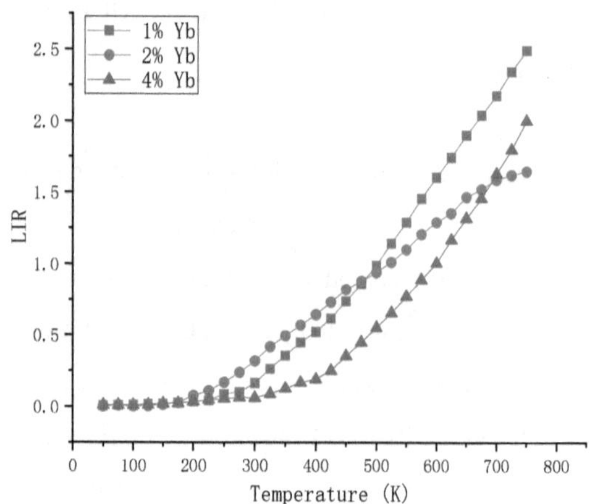

Figure 5. 5. LIR-T plots under irradiation of Argon Laser

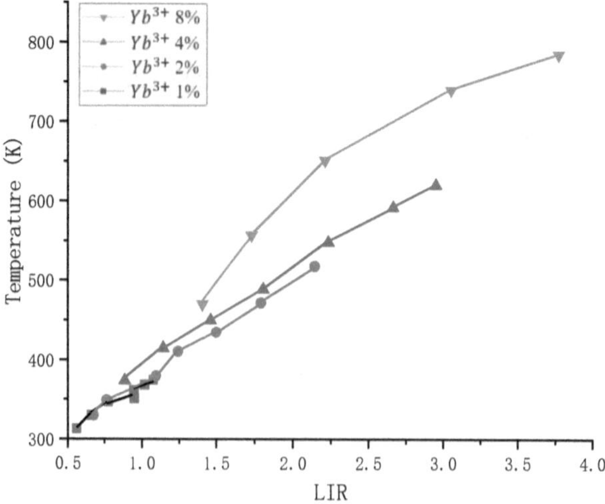

Figure 5. 6. LIR-T plots for samples SZErYb1, 2, 4 and 8 under irradiation of diode laser

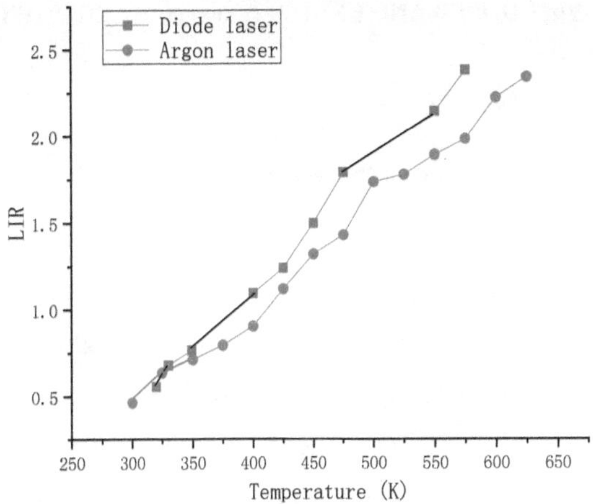

Figure 5. 7. LIR-T plots comparison under irradiation of Diode Laser and Argon Laser

5.4 CONCLUSION

We successfully achieve upconversion luminescence with the pellets of $SrZrO_3$, doped with Er^{3+} and Yb^{3+}, pumped by the 975nm diode laser in a vacuum. The luminescence light (green and red) comes from the electron energy level transition in Er^{3+} ions, and Yb^{3+} ions make this process possible. High laser power and Yb^{3+} concentration will increase the temperature in the sample, and this will cause the non-radiative loss and reduce the emission intensity. In the sample SZErYb8 with highest Yb^{3+} (8%), as the laser power reach to a certain threshold value (5.55W), the WL appeared. The experiments suggest that WL is due to thermal radiation, and chapter 6 will show more details about how it behaves and which parameters impact its appearance.

6. UNCONVENTIONAL PRODUCTION OF BRIGHT WHITE LIGHT EMISSION BY Er^{3+}-Yb^{3+} DOPED AND UNDOPED $SrZrO_3$ NANO POWDERS

6.1 INTRODUCTION

In Chapter 5, our experiments show the White Light (WL) emission phenomenon in the sample: Er^{3+}-Yb^{3+} co-doped $SrZrO_3$ pellets with a high Yb^{3+} concentration (8%) under a high irradiate laser power (5.55W). The WL emission from Nano Crystalline doped multiply with lanthanides have been revealed in the past decades [55-60]. With the excitation of a Near Infrared diode laser, (as the power exceeds a certain threshold value), a bright broad band light was observed, and the spectrum of it covered almost the whole visible region. Many researchers have found and reported this kind of phenomenon recently. Strek et al. have reported the anti-Stokes WL emission from $LiYbP_4O_{12}$ [57], $NdAlO_3$ [58] and $LiNdYbP_4O_{12}$ [59] nano crystals. J. Wang et al. also reported WL emission from lanthanide oxides such as Tb_2O_3, Sm_2O_3 and CeO_2 [55], and crystals like $Yb_3Al_5O_{12}$, $(Yb,Y)_2O_3$ [56]. Federico González et al. also found WL in $Y_4Zr_3YbO_{12}$ nanopowders [10]. The production of WL has also been reported by Gokhan Bilir during the past few years by using Nd doped and nominally undoped Y_2O_3 nano powders [57] and Cr^{3+} doped and nominally undoped Garnet Nano powders of $Y_3Al_5O_{12}$ and $Gd_3Ga_5O_{12}$ [63] etc.

$SrZrO_3$ is an excellent Nano-phosphors host for rare earth dopants due to its thermal and chemical stability, high melting point of 2873 K and strong luminescence properties. The lanthanide ions also have the attractive up-conversion luminescence ability that up-convert infrared light to the visible light, which interests the researchers a lot. Among them, Er^{3+} and Yb^{3+} ions are very good ion pair for applications of the up-conversion process because of their transition energy (10×10^{-3} cm^{-1}) in the $^4I_{15/2} \rightarrow {}^4I_{11/2}$ states of Er^{3+} and the $^2F_{7/2} \rightarrow {}^2F_{5/2}$ states of Yb^{3+} that has the similar amount of the 975 nm diode laser in our lab.

The conventional Nano-phosphor based lighting is widely used, and it is cheap and efficient. However, the luminescence spectrum of it doesn't cover the whole visible range light the thermalized black body radiation, so it's not very pleasant for human eyes (the best for human eyes is the sunlight). That is to say, the conventional artificial lighting still has a drawback, and the research for finding a new source of light that close to the sunshine will have a good prospect for practical applications. In this chapter, we are focusing on using $SrZrO_3$ nano powders doped and undoped with Er^{3+}, Yb^{3+} to obtain visible WL with a 975 nm infrared diode laser.

6.2 EXPERIMENT

The sample $SrZrO_3$ nano pellets and powers co-doped with 0 and 1% Er^{3+} and 1, 2, 4, 8% Yb^{3+} (SZErYb1, 2, 4, 8) were synthesized by *Professor Federico Garcia Gonzalez* in Universidad Autónoma Metropolitana in Mexico.

The emission spectra of WL were measured at room temperature (300K) and vacuum condition with a pressure of 0.001 mbar. The sample was pumped by a laser diode power supply LDI-820 with emitting wavelength at 975 nm and maximum output of 6.79 W, and the luminescence signal was directed to the entrance slit of a 1-meter McPherson monochromator, model 2051, which chopped at a frequency of

250 Hz before entering the slit, and detected by a Hamamatsu R1387 photomultiplier tube connected to a EG&G lock-in amplifier, model 5210, and recorded in the range from 400 nm to 800 nm by a computer. A 900 nm short-pass filter was used to block the signal from the laser. In addition, a Kolmar Technologies model KJSDP-1-JI/DC infrared detector was used to measure the spectrum in infrared part at a range of 800 nm to 1700 nm.

A HL-3 plus-CAL Calibrated light source was used to create spectra that were corrected for the responses of both detectors (photomultiplier and IR detector) and the monochromator.

A 50 MHz 4 Channels digital storage oscilloscope (DSO), model Rigol DS1054Z, was used to measure the rise and decay pattern of WL. Every 30 seconds after the light reaching its maximum intensity, a black shutter was used to cut sharply the pumping power, and the rise and decay pattern was recorded in the oscilloscope and saved with a memory disc. Moreover, a closed cycle Helium refrigerator was used to produce the environment temperature at a range of 50-800 K, and the temperature was controlled by a Lake Shore Cryotrons 331 Model temperature controller.

An Allied Scientific Pro ASP-MK350 model illuminance meter was also used to directly measure the emission spectrum of the WL accompanied with their CIE (International Commission of Illumination) coordinates, the CCT (Correlated Color Temperature), the CRI (Color Rendering Index) and illuminance. It could also be used to measure other commercial bulbs for comparison.

6.3 RESULTS AND DISCUSSION

As we discussed in last chapter, WL only appears in samples with a high Yb^{3+} concentration and a high laser power. However, after several test and investigation, we found that the production of WL in powders is much easier than in pellets, even in the samples with no Yb^{3+}. For the SZErYb8 sample, the pumping power of reaching clear luminescence for powders is much lower (0.22 W) than it for pellets (1.54 W); moreover, as the power reaches to 1.10 W, a very bright and clear yellowish WL appears, and the threshold value is much lower compared with the experiment results on pellets (5.55 W) in last chapter, where the WL was much harder to be produced as the laser almost pumped to its maximum power. The lower threshold of pumping power shows that the powders favor the emission of the luminescence and WL, comparing with the pellets. Figure 6.1 shows the comparison of WL emission intensity in pellets and powders of the SZErYb8 sample with the same condition. From the graph, we can notice that the emission signal from the powders is much stronger than the pellets. Thus, in this chapter, I will only focus on the behavior of WL emission in powders.

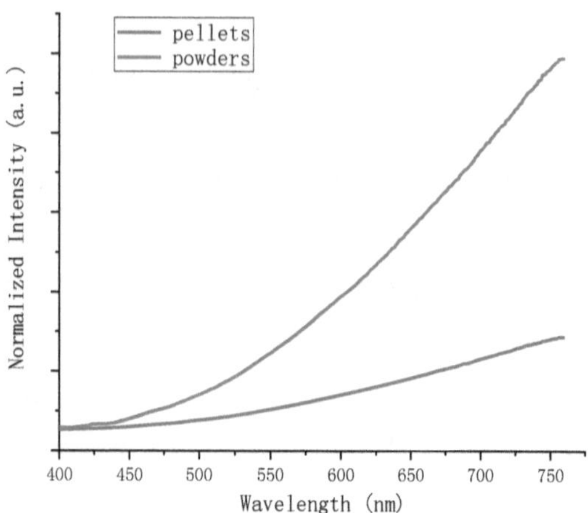

Figure 6. 1. WL spectra of sample SZErYb8 within pellets and powders, measured at the same (975 nm) excitation laser power.

6.3.1 Spectra of WL emission

Figure 6.2 (a) shows the WL spectrum taken between 400-800 nm wavelength range of the powders SZErYb8 (annealed in 1100C) corresponding to the highest (8%) Yb^{3+} concentration, under excitation at 975 nm with pumping power of 1.10 W to 6.79 W at a pressure of 0.1 mbar and room temperature. As we can see, the integrated intensity of WL emission increases as the power increases. Also, the emission intensity increases as the wavelength increases. Figure 6.2 (b) shows the comparison of WL spectrum as the pumping power rises and falls at the same environmental condition. The spectra are collected by increasing the power from 1.10 W to 6.79 W first, and then by decreasing the power back to 1.10 W. For the two emission intensities at the same power, the intensity when decreasing the laser power is reduced as compared when increasing. This fact indicates that the sample likely undergoes some modification after it irradiated by the laser.

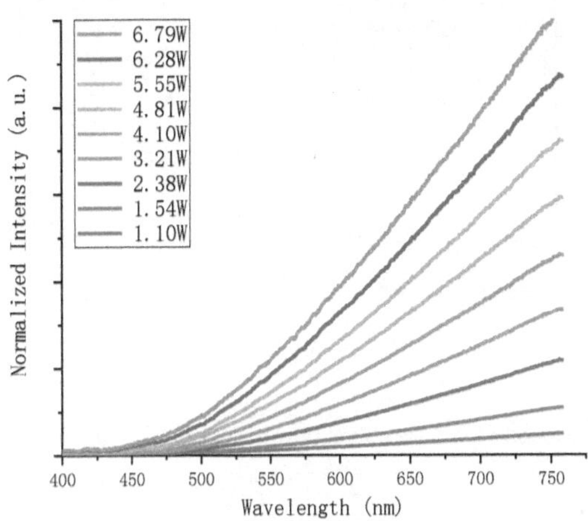

Figure 6. 2 (a). WL spectra of powder SZErYb8 at different excitation laser powers.

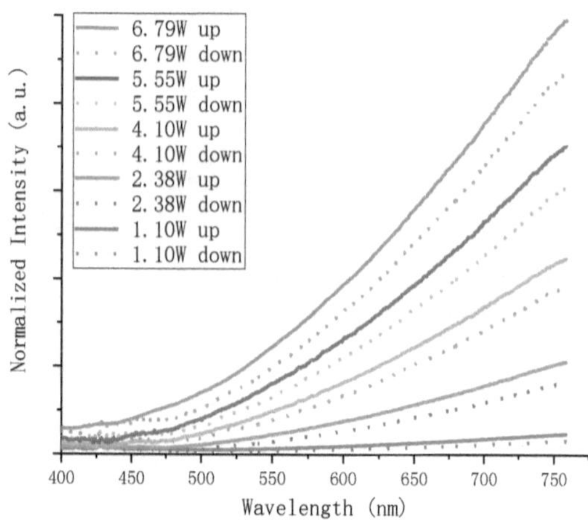

Figure 6. 2 (b). Comparison of WL spectra in powder SZErYb8 as the pumping power rise and decay.

Figure 6.3 shows the actual pictures of WL emission we took in the lab, each shows the emission of 4 difference concentrations of Yb^{3+}.

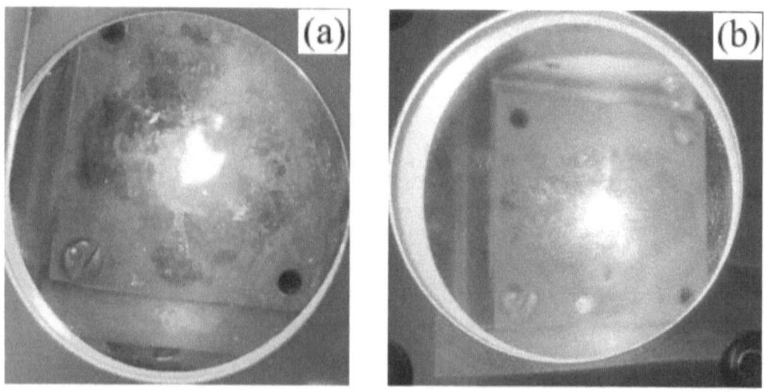

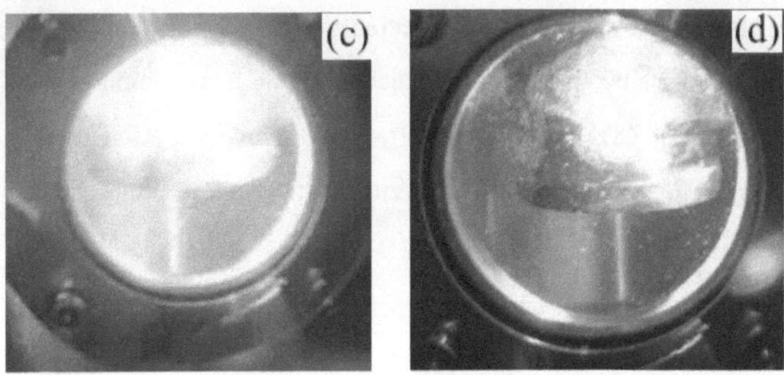

Figure 6. 3. WL emission of powders SZErYb1(a), 2(b), 4(c), 8(c).

Figure 6.4 shows the WL emission spectra in powder SZYb8 (without Er^{3+}, annealed in 1100C). Compared with SZErYb8, it doesn't show much difference in the spectra, and the intensity seems to be at the same magnitude within the same experimental condition. Therefore, the existence of Er^{3+} doesn't impact much of the WL production.

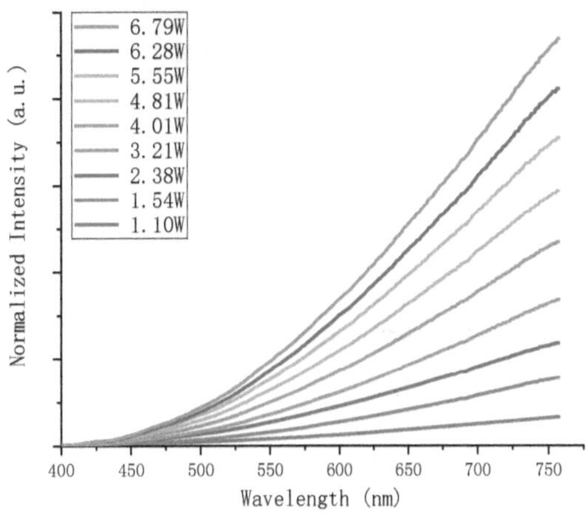

Figure 6. 4. WL spectra of powder SZYb8 at different excitation laser powers

Figure 6.5 shows the WL spectrum collected in the 400-800 nm wavelength range of the powders SZErYb0, 1, 2, 4, 8 (annealed in 1100C), corresponding to the lowest (0%) to the highest (8%) Yb^{3+} concentration, under excitation at 975 nm with pumping power of 5.55 W and various environment temperatures (100 K, 300 K and 500 K).

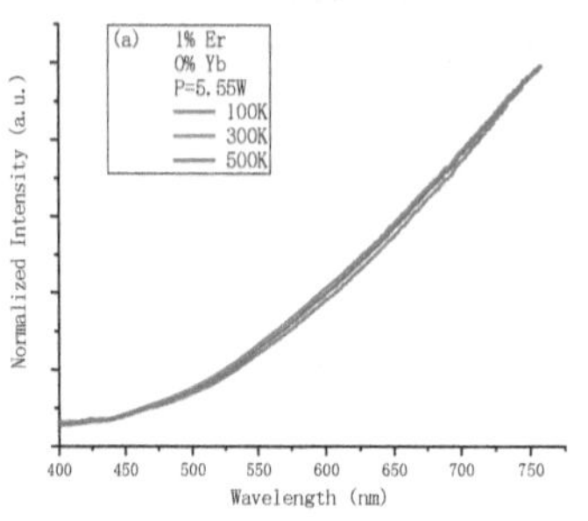

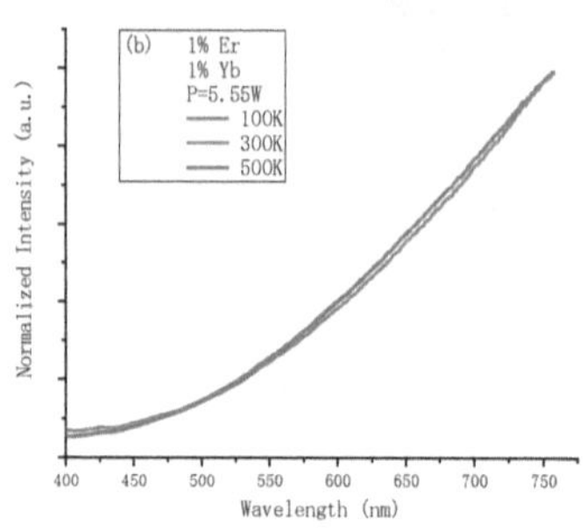

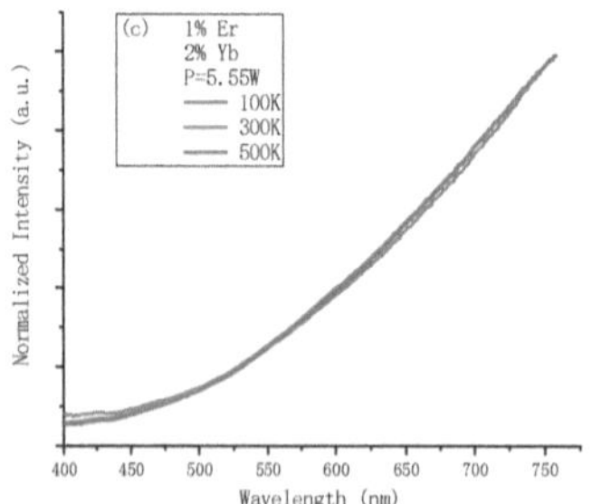

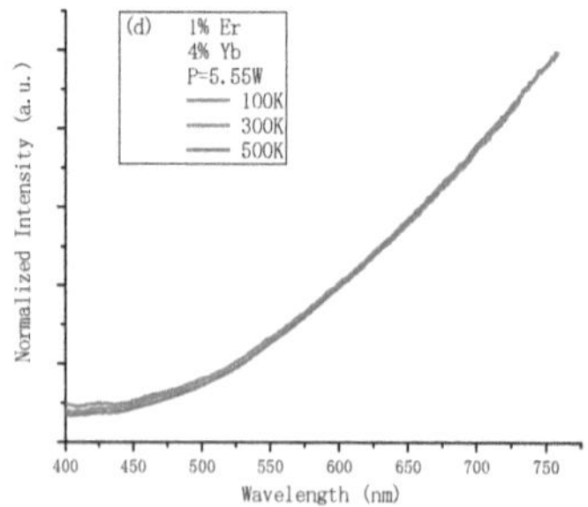

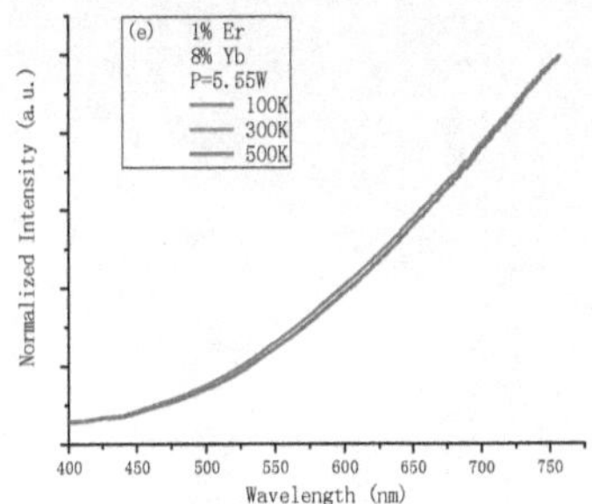

Figure 6. 5. Environment temperature dependence of WL spectra in powders (a) SZErYb0, (b) SZErYb1, (c) SZErYb2, (d) SZErYb4, (e) SZErYb8 under 975 nm excitation laser powers

From the figures above, obviously, the ambient temperature doesn't impact the appearance and general aspect much. However, the temperature at the emission center of the sample may be different depending on the concentration of the doped elements, the laser power etc., and it could be approximately calculated through Wien's Displacement law if we consider the system as a black body radiation model.

Figure 6.6 shows the spectra of the samples SZErYb0-8 measured by the illuminance meter under the laser power of 5.55W, and they show a good agreement with the corrected spectra from our experiment.

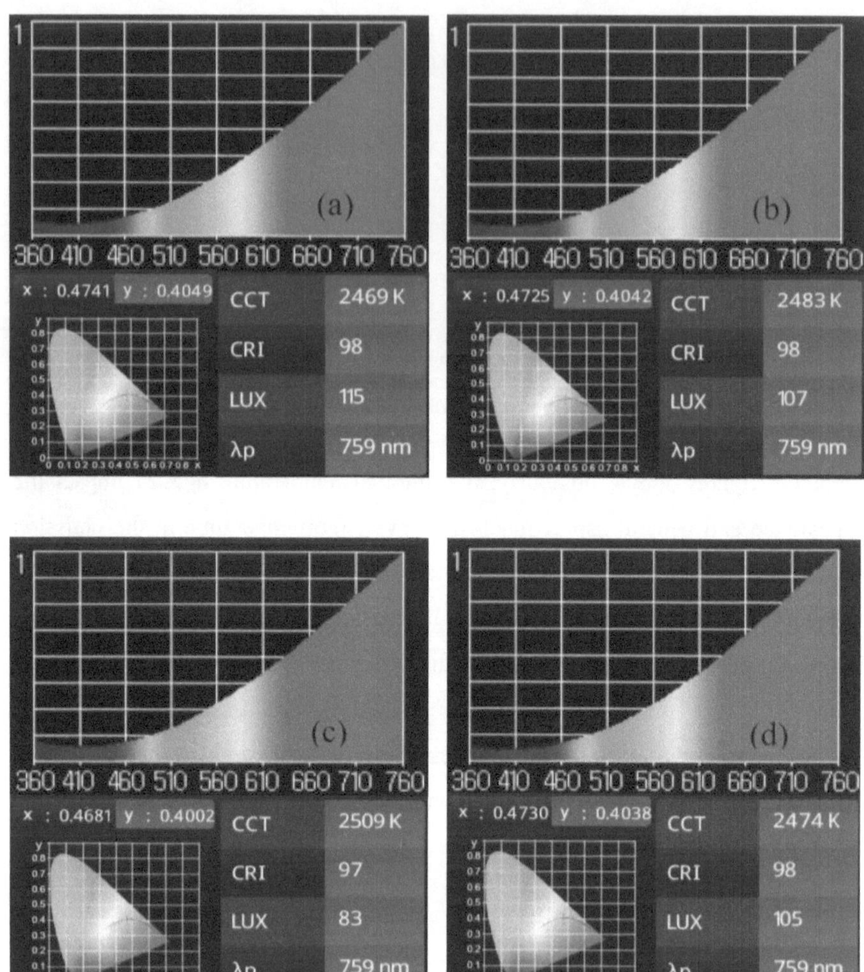

(a)
360 410 460 510 560 610 660 710 760
x : 0.4741 y : 0.4049
CCT 2469 K
CRI 98
LUX 115
λp 759 nm
(b)
360 410 460 510 560 610 660 710 760
x : 0.4725 y : 0.4042
CCT 2483 K
CRI 98
LUX 107
λp 759 nm
(c)
360 410 460 510 560 610 660 710 760
x : 0.4681 y : 0.4002
CCT 2509 K
CRI 97
LUX 83
λp 759 nm
(d)
360 410 460 510 560 610 660 710 760
x : 0.4730 y : 0.4038
CCT 2474 K
CRI 98
LUX 105
λp 759 nm

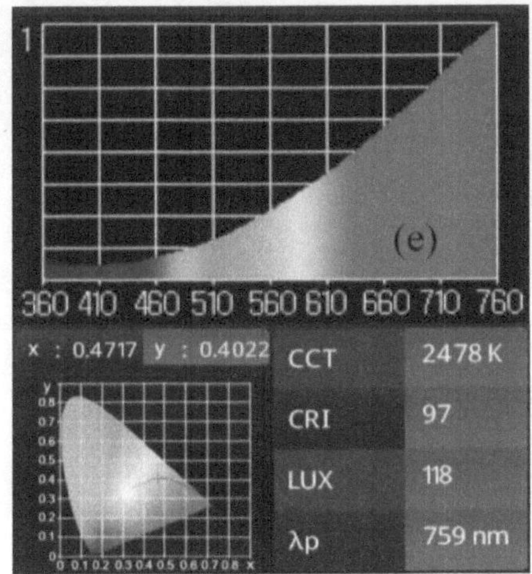

Figure 6. 6. WL spectra and CIE coordinates and CCT, CRI, LUX values measured by Illuminance Meter in powders (a) SZErYb0, (b) SZErYb1, (c) SZErYb2, (d) SZErYb4, (e) SZErYb8 under the same excitation laser power (5.55W)

The illuminance meter directly provided the International Commission on Illumination (CIE) coordinates of the WL emitted by the samples, and also shows the parameters of CCT (Correlated Color Temperature), CRI (Color Rendering Index), illuminance and the wavelength of peak. These parameters are also shown in the Table 6.1 below.

Table 6. 1 Summary of results by illuminance meter.

Sample	CIE	CCT	CRI	Efficiency
SZErYb0	x=0.474, y=0.405	2469K	98	20.7 lum/W
SZErYb1	x=0.471, y=0.404	2483K	98	19.3 lum/W
SZErYb2	x=0.468, y=0.400	2509K	97	15.0 lum/W
SZErYb4	x=0.473, y=0.402	2474K	98	18.9 lum/W
SZErYb8	x=0.472, y=0.402	2478K	97	21.3 lum/W

The CIE coordinates means chromaticity coordinates, and it reflects the hue and saturation of colors. CCT stands for the theoretical temperature of Planckian radiator whose chromaticity is nearest to that of the light source. CRI reflects the ability of a light source to reproduce the colors of various objects faithfully compared with a black-body light source. Illuminance, means the light flux received on unit area. Wavelength of peak gives the wavelength which has the maximum intensity in the range shown.

Figures 6.7 (a) and (b) show the power dependence of emission intensity in the powder SZErYb8 (1100 C) at 700 nm. Figure 6.7 (b) indicates that WL intensity varies with the pumping power P according to the law $I = P^n$, where n is an exponent which related to the number of photon transitions involved in the emission process [57, 64]. In our experiment, the exponent is found as 15.75 and 9.48 by linear fitting in two different regions.

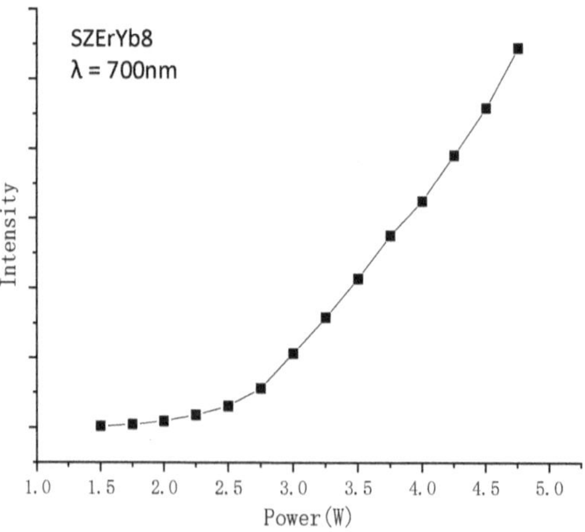

Figure 6. 7 (a) WL intensity variation with pumping power in powder SZErYb8

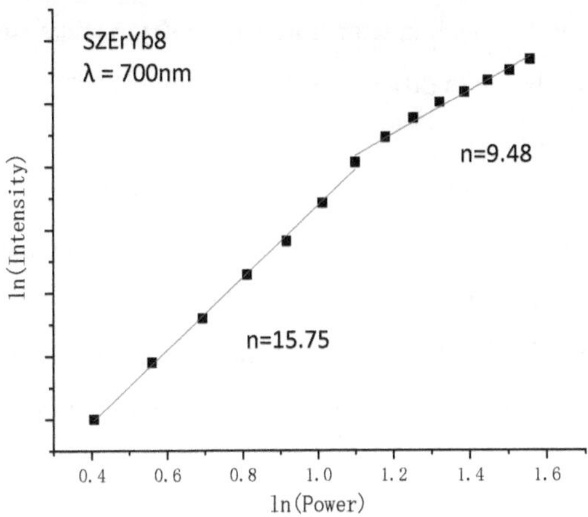

Figure 6. 7 (b) Ln (intensity) variation with Ln (power) with linear fit in powder SZErYb8

6.3.2 Threshold Dependence of WL emission

As we have already noticed, the environment temperature doesn't impact the appearance of WL. However, it could be affected by many other variables, such as the laser power, pressure, percentage composition of Er^{3+} and Yb^{3+}, the size of the nano crystal and so on. We used control variate method in the experiment to find the change mechanism of WL spectrum and the laser power threshold of WL appearance in different samples.

Concentration of Yb^{3+}:

We first focus on the impact of Yb^{3+} concentration. We set the environment temperature T = 300K, pressure P= 0.001mbar = 0.1Pa, and selected the samples which were at the same annealing temperature (1100C) to make sure that they have the same crystal size, but with different concentration of Yb^{3+}.

We investigated the threshold for WL appearance of powders SZErYb0, 1, 2, 4 and 8, corresponding to 0, 1, 2, 4 and 8% Yb^{3+} concentration respectively, and observed the variation of WL emission in each sample. The threshold value for each of them are 3.82 W, 3.42 W, 2.8 W, 1.96 W, 1.1 W for samples SZErYb0, 1, 2, 4 and 8, respectively. Figure 6.8 shows the comparison of the emission intensity from the 5 samples at the same excitation powers (2.5 W and 4.01 W).

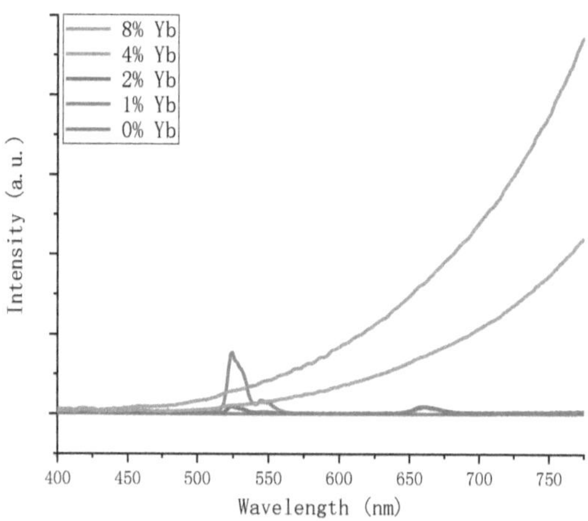

Figure 6. 8 (a). WL emission intensity of powders SZErYb0, 1, 2, 4, 8 under 2.5 W

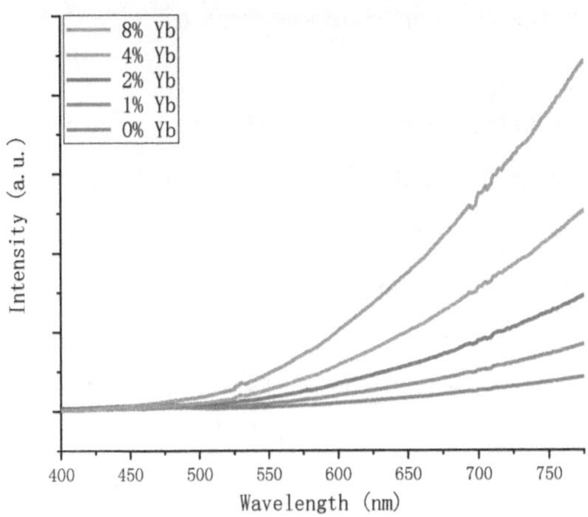

Figure 6. 8 (b). WL emission intensity of powders SZErYb0, 1, 2, 4, 8 under 4.01 W

As we can see from the Figure 6.8 (a), the sample SZErYb0 doesn't provide any light at the lower power (2.5W). This is the same as what we found in Chapter 5; the absence of Yb^{3+} disfavors the energy absorption in the samples from the 975 nm laser. Samples SZErYb1 and 2, same as pellets, both provide luminescence at the lower power, with the green emission at 530 nm and 550 nm, and red emission at 660 nm, corresponding to the Er^{3+} energy level transitions of excited states $^2H_{11/2}$/ $^4S_{3/2}$ and $^4F_{9/2}$ to the ground state $^4I_{15/2}$ respectively. The intensity of luminescence in sample SZErYb2 is lower than in sample SZErYb1. This is because as the power increases close to the threshold for providing WL, the non-radiative losses increase, and the luminescence intensity decreases sharply. This is a similar situation to the pellets SZErYb8 discussed in Chapter 5. Samples SZErYb4 and 8 provide WL emission at this power because it is over their threshold. In Figure 6.8

(b), all samples provide WL emission at high power (4.01 W). Also the emission intensity increases as the Yb^{3+}concentration increases.

As a conclusion, we find that the presence of Yb^{3+} favors the emission of WL. With more concentration of Yb^{3+}, the threshold value of the laser power is lower, and it is easier to produce WL. Table 6.2 summarizes the threshold values for getting WL in pellets and powders with different concentrations of Yb^{3+}.

Table 6. 2 Threshold for getting WL in samples with different Yb^{3+} concentration

Pellets (Yb^{3+}%)	0	1	2	4	8
Threshold (W)	No	No	No	No	5.55
Powders (Yb^{3+}%)	0	1	2	4	8
Threshold (W)	3.82	3.42	2.8	1.96	1.1

Nano particle annealing temperature

Second, we focus on the impact of the particle annealing temperature. We set the ambient temperature T = 300K, pressure P= 0.001mbar = 0.1Pa, and select the samples with the same Yb^{3+} concentration, but with different annealing temperature. A higher annealing temperature generally gives a larger particle size, but this was not confirmed for our samples. The samples are divided into 4 groups with 1%, 2%, 4% and 8% of Yb^{3+}. In each group, there are 2 samples with 2 different annealing temperature (900C and 1100C) for comparison. The samples are irradiated under 975 nm laser from low power to high power until WL appeared. Figure 6.9 shows the emission spectra for samples at 5.55W.

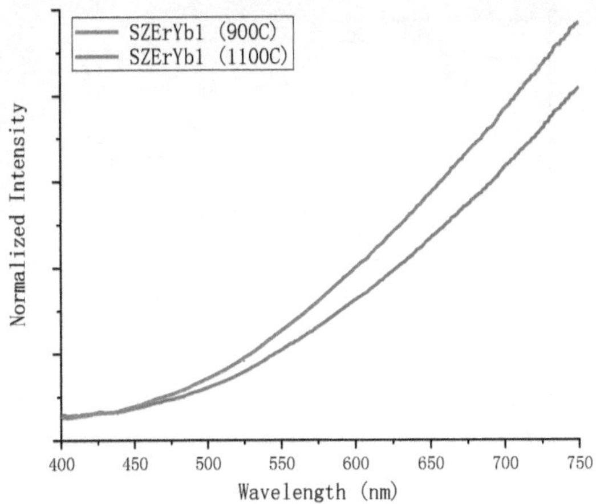

Figure 6. 9 (a). WL emission intensity of powders SZErYb1 (900C and 1100C)

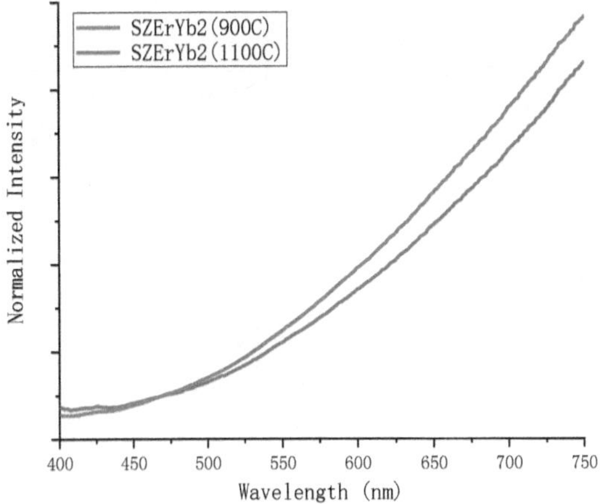

Figure 6. 9 (b). WL emission intensity of powders SZErYb2 (900C and 1100C)

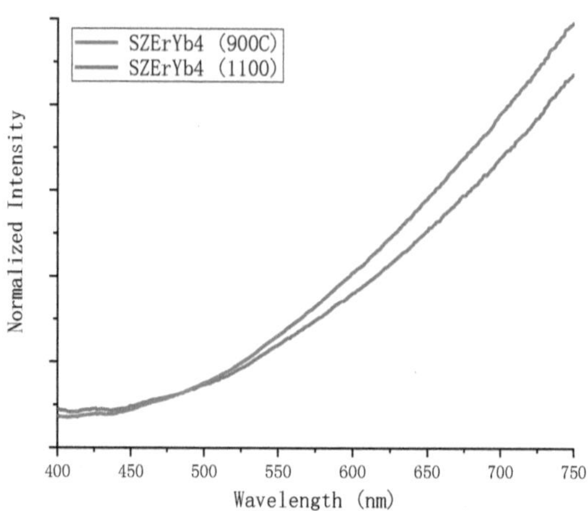

Figure 6. 9 (c). WL emission intensity of powders SZErYb4 (900C and 1100C)

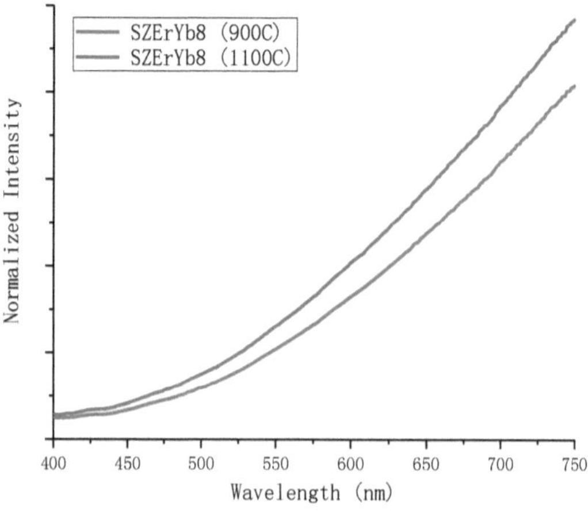

Figure 6. 9 (d). WL emission intensity of powders SZErYb8 (900C and 1100C)

The WL intensity for both samples are quite close within error, thus, the nano particle annealing temperature doesn't impact the emission intensity. The threshold for both samples are slightly different, lower annealing temperature gives a lower value. As a conclusion, the lower annealing temperature favors the WL emission. Table 6.3 summarizes the results.

Table 6. 3 Threshold for getting WL in samples with different annealing temperature

Yb^{3+} %	0	1	2	3	4
Threshold 1100C (W)	3.82	3.42	2.8	1.96	1.1
Threshold 900C (W)	3.65	3.1	2.59	1.74	0.89

Pressure Dependence

Previous experiments indicate that it is easy to generate WL in low pressure. We also tried the same measurements in the atmosphere, but didn't see any light at all. So the environment pressure also impacts the WL production. Now we focus on the pressure dependence, and study the corresponding variation of WL spectra.

We set the laser power to be 6.62 W and the ambient temperature T = 300K to make sure that we could get as much WL as we can. First, we decrease the pressure to 0.001 mbar, and record the intensity of WL emission at 700 nm; then gradually increase the pressure by 10 times, and record the corresponding intensity. The WL variation diagrams and Intensity–Pressure diagrams of sample SZErYb8 (1100C) and SZErYb0 (900C) are shown in Figure 6.10 and 6.11.

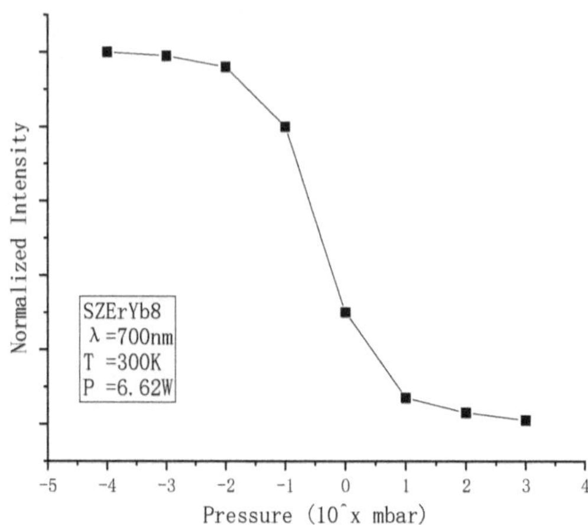

Figure 6. 10. WL intensity variation with pressure in powder SZErYb8 (annealed at 1100C)

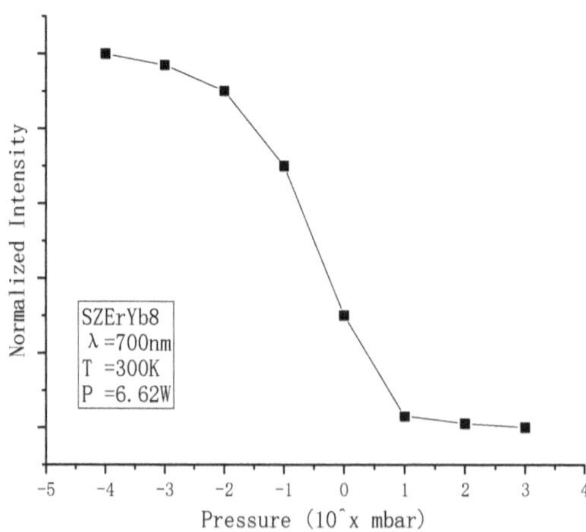

Figure 6. 11. WL intensity variation with pressure in powder SZErYb0 (annealed at 900C)

From the figures 6.10 and 6.11, we see that both samples reveal the same conclusion that the WL intensity falls down rapidly with the increasing of pressure. Moreover, for sample SZErYb8 (1100C), the threshold of laser power to obtain WL under 0.02 mbar was found to be above 4.01 W, much higher compared to what we found before under 0,001 mbar (1.1 W). For sample SZErYb0 (900C), the threshold under 0.02 mbar was above 6.28 W, also much higher than the condition under 0,001 mbar (3.65 W).

Based on the experimental results, we conclude that the existence of Er^{3+} and the environment temperature doesn't impact the WL much, neither the emission intensity nor the threshold. The annealing temperature didn't impact the emission intensity, but lower annealing temperature favors the threshold a little bit. High concentration of Yb^{3+} and low pressure favor both emission intensity and threshold, however, the existence of Yb^{3+} is not a necessary condition to provide WL. Thus, the most efficient WL comes from the SZErYb8 (900C) at a low pressure.

6.3.3 Rise pattern and decay pattern

In order to get a deeper understanding of the dynamic characteristics of the WL, the rise and decay patterns following a sudden excitation and a sudden interruption of the infrared radiation, were measured. Both patterns could be affected by experimental variables like the laser power, pressure, temperature and Yb^{3+} amount.

Figure 6.12 shows the decay patterns of WL in powder SZErYb8 (1100C) at different environmental temperature. As we can see from the graph, the decay patterns are very sensitive to the ambient temperature. When the temperature goes higher, the decay times becomes longer.

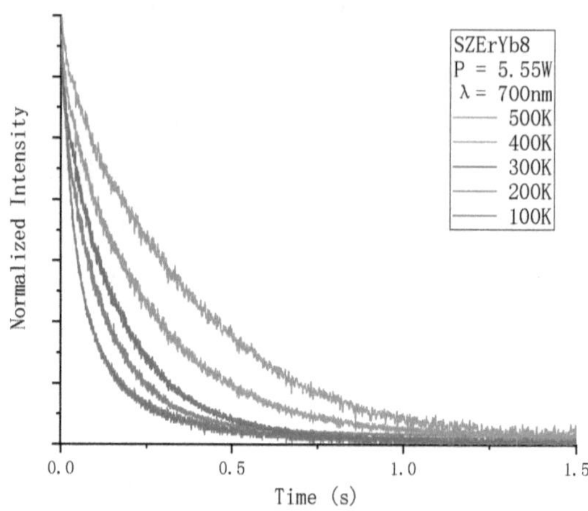

Figure 6. 12. Dependence of decay patterns on temperature in powder SZErYb8

Figures 6.13 and 6.14 shows decay patterns and rise patterns in powder SZErYb8 (1100C) in different pressure. Both decay and rise patterns are sensitive to the pressure. When the environment pressure goes higher, the decay time is shorter while the rise time is longer.

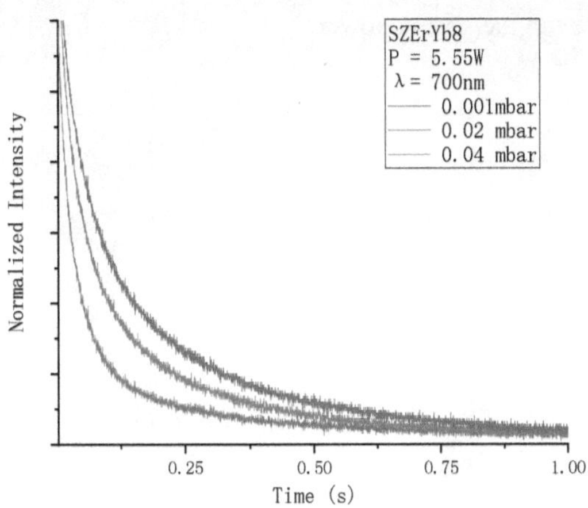

Figure 6. 13. Dependence of decay patterns on pressure in powder SZErYb8

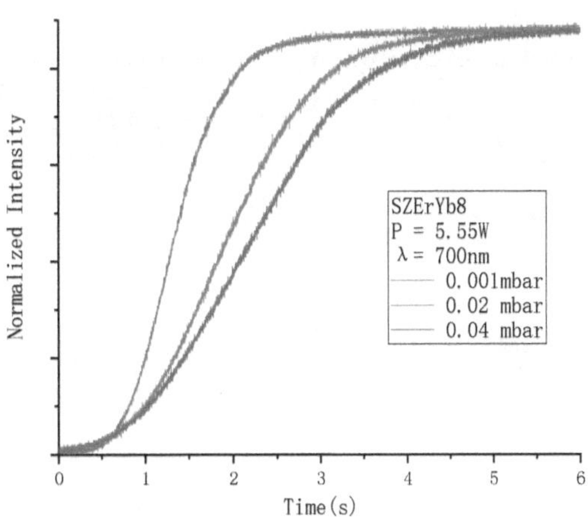

Figure 6. 14. Dependence of rise patterns on pressure in powder SZErYb8

Figures 6.15 and 6.16 shows decay patterns and rise patterns in powder SZErYb8 (1100C) under different excitation laser power. The decay patterns are not very sensitive with the variation of power, while the rise patterns are very sensitive, becoming shorter with increasing power.

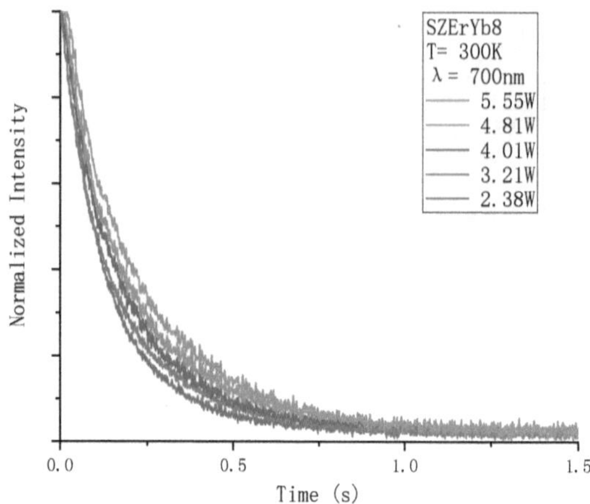

Figure 6. 15. Dependence of decay patterns on excitation power in powder SZErYb8

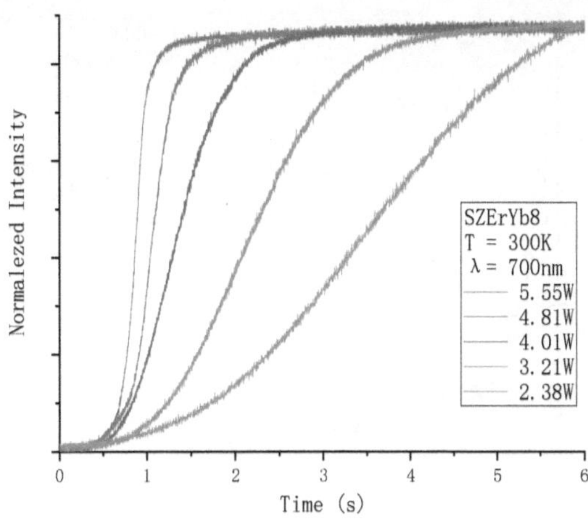

Figure 6. 16. Dependence of rise patterns on excitation power in powder SZErYb8

Figures 6.17 and 6.18 shows decay and rise patterns in samples SZErYb0-8 with different Yb^{3+} concentration, from 0 to 8%, within the same experiment condition. The decay patterns are not sensitive with the Yb^{3+} concentration, which gives an opinion that the WL decay is an host-dependent effect. The rise patterns are sensitive to Yb^{3+} concentration. In particular, they show that the rise time has an obvious delay which depends on the Yb^{3+} concentration before the WL appears, and the delay time becomes longer with decreasing Yb^{3+} concentration.

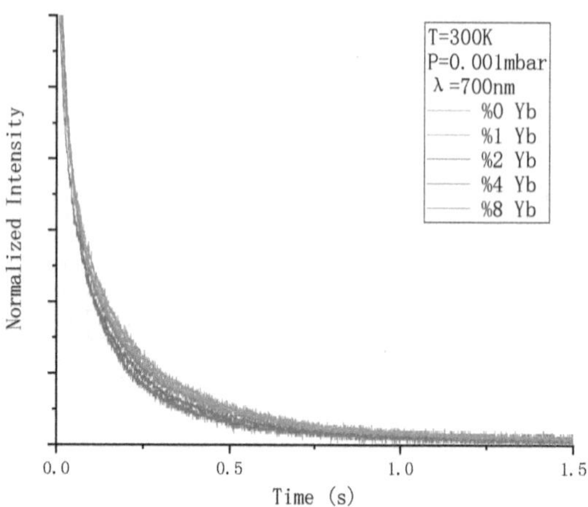

Figure 6. 17. Dependence of decay patterns on different Yb^{3+} concentration

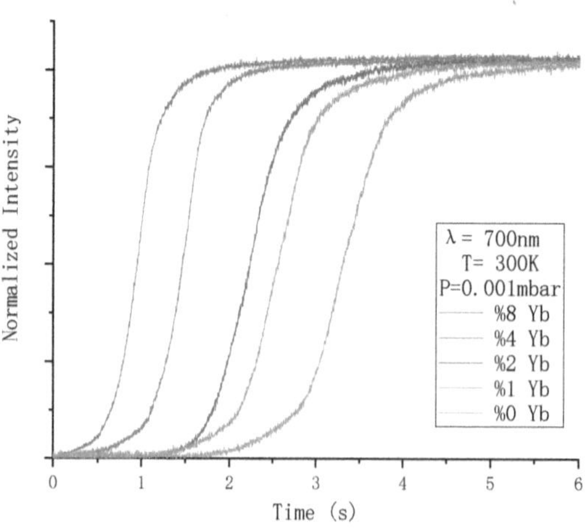

Figure 6. 18. Dependence of rise patterns on different Yb^{3+} concentration

Figure 6.19 shows decay patterns in samples SZErYb8 at different wavelength of the WL emission (400 nm, 550 nm and 700 nm) under 5.55W. The decay patterns do not change much with the variation of wavelength.

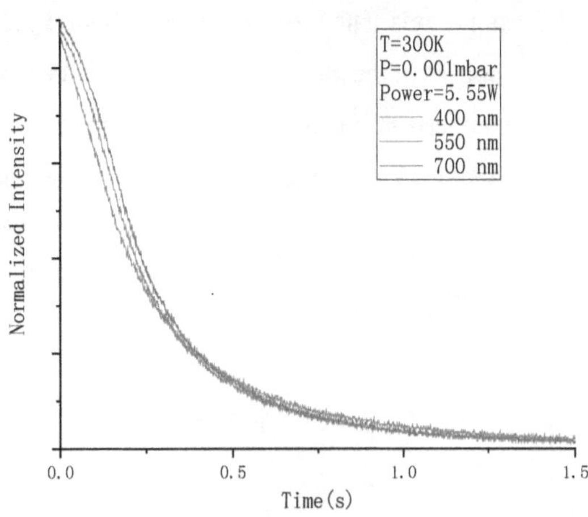

Figure 6. 19. Dependence of decay patterns at different emission wavelengths

6.3.4 WL Emission Characterization in Infrared part

Chapter 5 gives us an enlightenment that the original of the WL might be thermal radiation. Based on this hypothesis, we would like to know what the emission spectrum in infrared part looks like and whether the actual temperature matches with the theoretical value. A Kolmar Technologies model KJSDP-1-JI/DC infrared detector was used to measure the WL spectrum at a range of 800 nm to 1700 nm. With the combination of the visible range and the infrared range, Figure 6.20 shows the spectra of sample SZYb0 (1100C) from 400 nm to 1600 nm at three different powers (4.81, 5.55 and 6.28 W), and Figure 6.21 shows the spectra of sample SZYb8 (1100C) under the same conditions. The spectra shown in Figures 6.20 and 6.21 have been corrected for the responses of the monochromator and of the two detectors used in the experiments. For comparison, the theoretical emission spectrum (T=3000 K) based on Planck's law in the black body radiation model is shown in Figure 6.22.

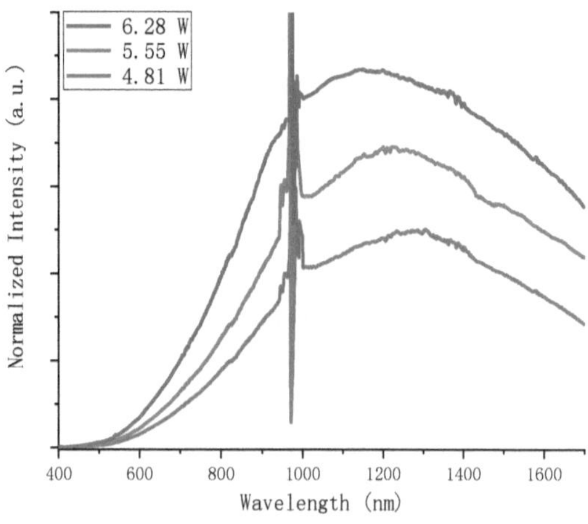

Figure 6. 20. WL spectra of powder SZYb0 at whole range

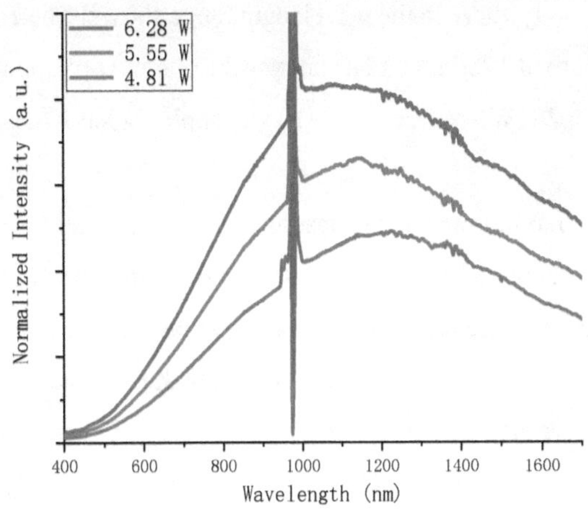

Figure 6. 21. WL spectra of powder SZYb8 at whole range

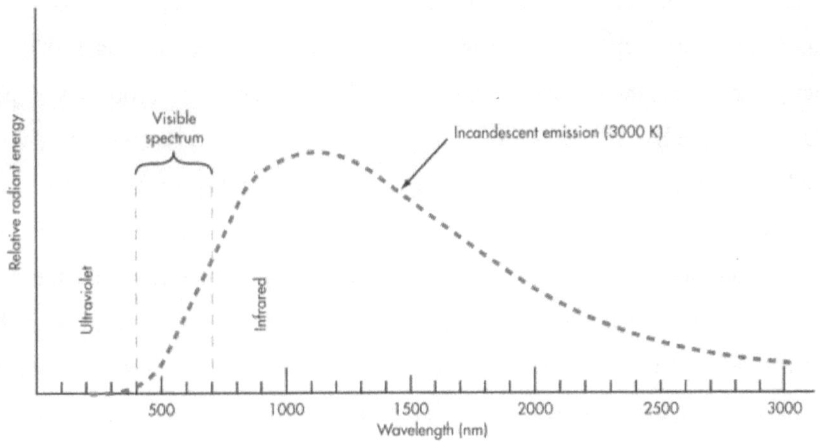

Figure 6. 22. theoretical black body emission spectrum (T=3000 K)

In general, the shape of the combined spectra from our experiment match the general shape of the black-body curve in the wavelength range shown. The emission maxima of sample SZYb0 are at 1286 nm, 1219 nm and 1137 nm, corresponding to the 4.81 W, 5.55 W and 6.28 W, respectively. If we apply Wien's Displacement law:

$$\lambda_m \cdot T = b$$

λ_m is the emission maximum, T is the absolute temperature, and b = 0.002897m·K. The corresponding temperatures are 2253 K, 2377 K and 2548 K, and these values are in the same range as the temperature (2469K) that we measured directly by the illuminance meter. We observed similar results for the sample SZYb8, where the emission maxima are at 1220 nm, 1142 nm and 1089 nm which correspond to calculated temperatures of 2375 K, 2536 K and 2660 K, respectively, which are also in the same range as the temperature (2509 K) from the illuminance meter.

On the other hand, reaching such a high temperature is likely to cause the sample to start melting since the melting point of the $SrZrO_3$ nano crystal is just 2873 K. Figure 6.23 shows the surface of the powder SZYb8 under a microscope. As we can see, the sample has already started melting at the excitation area after a half-hour irradiation under the power of 6.28 W. The melting effect happens the most at the center of irradiation, and decreases gradually in all directions. This phenomenon also demonstrates an evidence that the irradiation temperature in the sample is more or less at 2873 K, and is consistent with the values that we calculated through the Wien's Displacement law. This result provides further evidence that the origin of the WL is from thermal radiation. However, the details of the mechanism regarding how the transition to WL occurs is still a mystery, and requires further research.

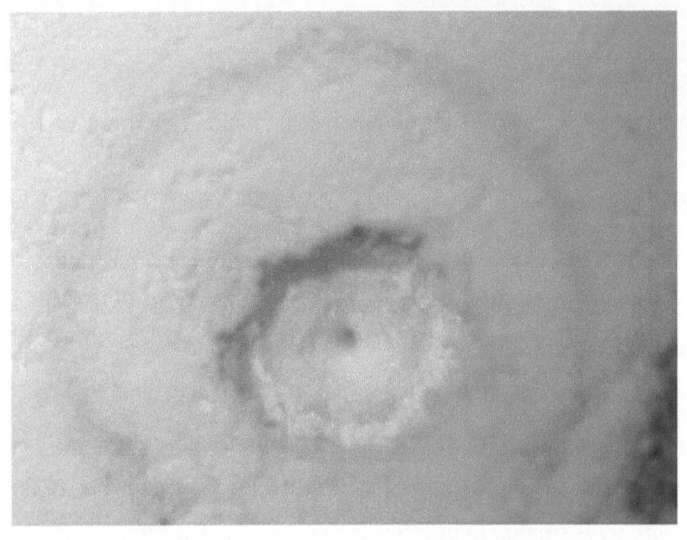

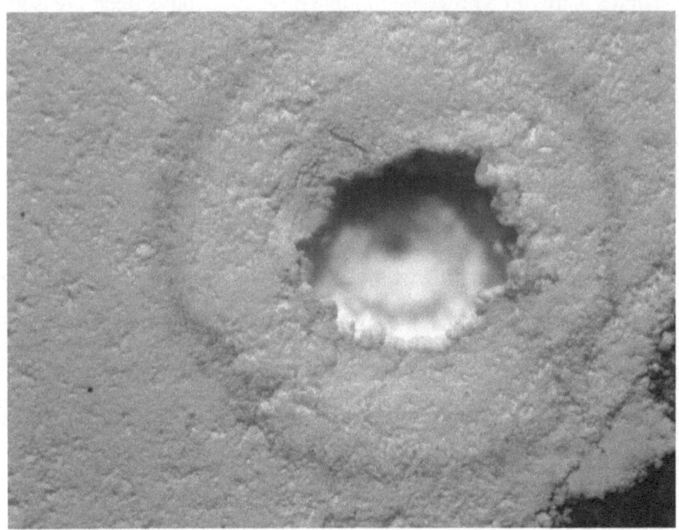

Figure 6. 23. Microscope pictures of the powder SZYb8 surface

6.4 CONCLUSIONS

Based on the experimental results above, we can make some basic observations as follows:

1. The yellowish Anti-Stokes WL emission is a threshold phenomenon with samples under the irradiation of a focused infrared laser diode, and the threshold value relies on different parameters.

2. It is much easier to generate WL in powders than in pellets within the same environment as the threshold in powders is much lower. This is due to the fact that heat conductivity is smaller in powders than in pellets. The thermal contact between the nanoparticles is low in a powder, and in a pellet they have been pressed together, so that the thermal contact between particles is much higher. The reduced heat conduction limits the ability of the heat to move away from the laser focus into the surrounding solid, allowing for faster heating of the sample. This is consistent with the white light being thermal in origin.

3. The Er^{3+} dopant is the main factor for generating green and red luminescence based on the energy transitions in different energy states for low powers. WL emission is obtained for power levels above the threshold. The existence of Er^{3+} doesn't influence the WL emission intensity, spectral shape, or threshold. So the white emission has a totally different mechanism than the Er^{3+} emission.

4. The Yb^{3+} dopant favors the WL generation as well as the emission intensity. However, the existence of Yb^{3+} is not necessary to induce this process. Moreover,

the spectral distribution is also independent on the Yb^{3+} concentration. This signifies that the mechanism of WL only depends on the host materials.

5. The nano particle annealing temperature didn't significantly impact the emission intensity, but a lower annealing temperature slightly favors the threshold. The particles annealed in a lower temperature likely have a lower heat conductivity which allows greater heating at the laser focus, but the effect was not very significant.

6. Based on our experiment results, the temperature of the sample at the focus is in the range 2400K-2700K which is much higher than the ambient (Dewar's) temperature (100K-500K). So the ambient temperature doesn't have much impact on the emission process.

7. The ambient pressure has a great influence on WL emission: low pressure favors both higher intensity and lower threshold. This is due to the lower thermal conductivity in vacuum. At low pressure, the nano-particles are more isolated from the surrounding gas, and can easily reach higher temperatures that favors the light emission [60]. At high pressures the surrounding gas can carry the heat away from the sample more rapidly. When there is gas in the sample chamber, the air molecules can collide with the sample, and scattering effects can cause heating dissipation. If the sample is hot, the molecules will carry energy away from the hot sample, cooling it.

8. The dynamical expression of the WL emission is characterized by the rise and decay patterns which follow a sudden beginning or ending of the input infrared radiation. The rise patterns are sensitive to the experimental variables of pressure

(see Figure 6.14) and laser power (see Figure 6.16), while the decay patterns are sensitive to the ambient temperature (see Figure 6.12) and pressure (see Figure 6.13). The decay patterns are not sensitive to the Yb^{3+} concentration (see Figure 6.17). The faster rise time (see Figure 6.18) for higher concentration of Yb^{3+} indicates that the Yb^{3+} acts as an absorber of the laser energy.

9. The observed spectral distributions (including visible range and infrared range) for SZYb0 for three different power levels are shown in Figure 6.20. The observed spectral distributions for SZYb8 for the same three power levels are shown in Figure 6.21. In each case, the observed distributions are close to the theoretical black body emission curves (see Figure 6.22) for the corresponding temperatures. The melting phenomenon on the sample surface (see Figure 6.23) indicates that in the area near the laser focus the melting temperature was reached. The maximum temperature is at the center, and the temperature in the nearby area decreases as we move away from the center. Thus, it is possible to consider the observed curves in our experiment as a superposition of a number of black body curves with different temperatures, arising from the distributed population of the nano particles.

7. CONCLUSION

Our experiments show that the emission of luminescence and WL can be induced by $SrZrO_3$ nano-particles which are doped or un-doped with Er^{3+}-Yb^{3+} ions. The green and red luminescence are generated by the Er^{3+} energy level transition at a low laser power, while the WL is generated by the host material at a high power. Many basic parameters can affect the threshold and intensity of WL emission, like Yb^{3+} concentration, sample temperature, ambient pressure and nano particle size. Based on our results, the temperature at the irradiation point is the determinant of such emission, and the WL is induced by thermal radiation. The irradiation temperature, measured by the illuminance meter, is at a range of 2300K-2700K, and matches with the calculated temperatures by Wien's Displacement law from the combined spectra. The melting phenomena on the surface of sample also confirmed the fact that the temperature may reach such a high value. Thus, it is possible to consider our sample under high power excitation as a black body emitter.

The WL has been studied by a few groups in the past decade. Strek et al. reported their observation of WL in $LiYbP_4O_{12}$ nanocrystals. They believed that the origin of the light emission is associated with emission of Yb^{3+} charge transfer (CT) cluster, and they doubt the hypothesis that the origin of this emission is black body radiation [57, 58, 64]. However, our results show that the Yb^{3+} dopant ion is not necessary for WL emission. If it is present, it acts like an absorber that reduces the WL threshold power level and enhances the intensity. The WL is a host-dependent phenomenon, where the laser irradiation heats up the host to a high temperature. In addition, the measured temperature and the fact that the sample melted also confirmed the inference of black body radiation. Pervious experiments by González et al. [10] and Bilir et al. [60, 63] also interpreted the WL emission as thermal radiation. We have repeated many of their experiments in a new sample ($SrZrO_3$) and obtained similar results. However, we have extended the wavelength range of the observation to include both the visible and the near IR regions which have shown decisively that the spectra resemble very closely black-body emission.

The white light sources are pleasant for human eyes rather than the commercial fluorescent light, and thus have huge application potential. We believe that with the continuing study on this topic people will have this kind of new technology for daily life in the future.

REFERENCES

[1] Pang, T., Cao, W., Xing, M., Feng, W., & Xu, S. (2010). Blue and white upconversion emissions of rare-earth ions-doped oxyfluoride phosphors. *Physica B: Condensed Matter, 405*(9), 2216-2219. doi:10.1016/j.physb.2010.02.013

[2] Erdem, M., Erguzel, O., Ekmekci, M. K., Orucu, H., Cinkaya, H., Genc, S., . . . Bartolo, B. D. (2015). Bright white up-conversion emission from sol–gel derived Yb3+/Er3+/Tm3+: Y2SiO5 nanocrystalline powders. *Ceramics International, 41*(10), 12805-12810. doi:10.1016/j.ceramint.2015.06.116

[3] Passuello, T., Piccinelli, F., Pedroni, M., Bettinelli, M., Mangiarini, F., Naccache, R., . . . Speghini, A. (2011). White light upconversion of nanocrystalline Er/Tm/Yb doped tetragonal Gd4O3F6. *Optical Materials, 33*(4), 643-646. doi:10.1016/j.optmat.2010.11.017

[4] Cesaria, M., Collins, J., & Bartolo, B. D. (2016). On the efficient warm white-light emission from nano-sized Y2O3. *Journal of Luminescence, 169*, 574-580. doi:10.1016/j.jlumin.2015.08.017

[5] Anh, T., Minh, L. Q., Vu, N., Huong, T. T., Huong, N. T., Barthou, C., & Strek, W. (2003). Nanomaterials containing rare-earth ions Tb, Eu, Er and Yb: Preparation, optical properties and application potential. *Journal of Luminescence, 102-103*, 391-394. doi:10.1016/s0022-2313(02)00531-8

[6] Hehlen, M. P., Frei, G., & Güdel, H. U. (1994). Dynamics of infrared-to-visible upconversion inCs3Lu2Br9:1%Er3+. *Physical Review B, 50*(22), 16264-16273. doi:10.1103/physrevb.50.16264

[7] Riedener, T., Egger, P., Hulliger, J., & Güdel, H. U. (1997). Upconversion mechanisms inEr3+-dopedBa2YCl7. *Physical Review B, 56*(4), 1800-1808. doi:10.1103/physrevb.56.1800

[8] Balda, R., Garcia-Adeva, A. J., Voda, M., & Fernández, J. (2004). Upconversion processes inEr3+-dopedKPb2Cl5. *Physical Review B, 69*(20). doi:10.1103/physrevb.69.205203

[9] Debasu, M. L., Ananias, D., Pastoriza-Santos, I., Liz-Marzán, L. M., Rocha, J., & Carlos, L. D. (2013). All-In-One Optical Heater-Thermometer Nanoplatform Operative From 300 to 2000 K Based on Er3+Emission and Blackbody Radiation. *Advanced Materials, 25*(35), 4868-4874. doi:10.1002/adma.201300892

[10] González, F., Khadka, R., López-Juárez, R., Collins, J., & Bartolo, B. D. (2018). Emission of white-light in cubic Y4Zr3O12:Yb3+ induced by a continuous infrared laser. *Journal of Luminescence, 198*, 320-326. doi:10.1016/j.jlumin.2018.02.053

[11] Lei, Y., Song, H., Yang, L., Yu, L., Liu, Z., Pan, G., . . . Fan, L. (2005). Upconversion luminescence, intensity saturation effect, and thermal effect in Gd2O3:Er3,Yb3+ nanowires. *The Journal of Chemical Physics, 123*(17), 174710. doi:10.1063/1.2087487

[12] Bhargava, R. N., Gallagher, D., Hong, X., & Nurmikko, A. (1994). Optical properties of manganese-doped nanocrystals of ZnS. *Physical Review Letters, 72*(3), 416-419. doi:10.1103/physrevlett.72.416

[13] Okamoto, S., & Masumoto, Y. (1997). Correlation betweenCu+-ion instability and persistent spectral hole-burning phenomena of CuCl nanocrystals. *Physical Review B, 56*(24), 15729-15733. doi:10.1103/physrevb.56.15729

[14] Masumoto, Y., Kawabata, K., & Kawazoe, T. (1995). Quantum size effect and persistent hole burning of CuI nanocrystals. *Physical Review B, 52*(11), 7834-7837. doi:10.1103/physrevb.52.7834

[15] Huang, J., Yang, Y., Xue, S., Yang, B., Liu, S., & Shen, J. (1997). Photoluminescence and electroluminescence of ZnS:Cu nanocrystals in polymeric networks. *Applied Physics Letters, 70*(18), 2335-2337. doi:10.1063/1.118866

[16] Riwotzki, K., & Haase, M. (1998). Wet-Chemical Synthesis of Doped Colloidal Nanoparticles: YVO4:Ln (Ln = Eu, Sm, Dy). *The Journal of Physical Chemistry B, 102*(50), 10129-10135. doi:10.1021/jp982293c

[17] Ihara, M., Igarashi, T., Kusunoki, T., & Ohno, K. (2000). Preparation and Characterization of Rare Earth Activators Doped Nanocrystal Phosphors. *Journal of The Electrochemical Society, 147*(6), 2355. doi:10.1149/1.1393536

[18] Wei, Z., Sun, L., Liao, C., Yan, C., & Huang, S. (2002). Fluorescence intensity and color purity improvement in nanosized YBO3:Eu. *Applied Physics Letters, 80*(8), 1447-1449. doi:10.1063/1.1452787

[19] Song, H., Sun, B., Wang, T., Lu, S., Yang, L., Chen, B., . . . Kong, X. (2004). Three-photon upconversion luminescence phenomenon for the green levels in Er3+/Yb3+ codoped cubic nanocrystalline yttria. *Solid State Communications, 132*(6), 409-413. doi:10.1016/j.ssc.2004.07.044

[20] Vali, R. (2008). Band structure and dielectric properties of orthorhombic SrZrO3. *Solid State Communications, 145*(9-10), 497-501. doi:10.1016/j.ssc.2007.12.009

[21] Terki, R., Feraoun, H., Bertrand, G., & Aourag, H. (2005). Full potential calculation of structural, elastic and electronic properties of BaZrO3 and SrZrO3. *Physica Status Solidi (b), 242*(5), 1054-1062. doi:10.1002/pssb.200402142

[22] Matsuda, T., Yamanaka, S., Kurosaki, K., & Kobayashi, S. (2003). High temperature phase transitions of SrZrO3. *Journal of Alloys and Compounds, 351*(1-2), 43-46. doi:10.1016/s0925-8388(02)01068-x

[23] Kennedy, B. J., Howard, C. J., & Chakoumakos, B. C. (1999). High-temperature phase transitions inSrZrO3. *Physical Review B, 59*(6), 4023-4027. doi:10.1103/physrevb.59.4023

[24] Petit, P. E., Guyot, F., & Farges, F. (1997). Thermal Evolution of BaZrO3and SrZrO3Perovskites from 4 K to 773 K : An EXAFS Study at Zr-K Edge. *Le Journal De Physique IV, 7*(C2). doi:10.1051/jp4:19972138

[25] Evarestov, R. A., Bandura, A. V., Alexandrov, V. E., & Kotomin, E. A. (2005). DFT LCAO and plane wave calculations of SrZrO3. *Physica Status Solidi (b), 242*(2). doi:10.1002/pssb.200409085

[26] Meza, O., Diaz-Torres, L., Salas, P., Rosa, E. D., & Solis., D. (2010). Color tunability of the upconversion emission in Er–Yb doped the wide band gap nanophosphors ZrO2 and Y2O3. *Materials Science and Engineering: B, 174*(1-3), 177-181. doi:10.1016/j.mseb.2010.03.015

[27] Shang, Q., Yu, H., Kong, X., Wang, H., Wang, X., Sun, Y., . . . Zeng, Q. (2008). Green and red up-conversion emissions of Er3+–Yb3+ Co-doped TiO2 nanocrystals prepared by sol–gel method. *Journal of Luminescence, 128*(7), 1211-1216. doi:10.1016/j.jlumin.2007.11.097

[28] Gao, G., Busko, D., Kauffmann-Weiss, S., Turshatov, A., Howard, I. A., & Richards, B. S. (2017). Finely-tuned NIR-to-visible up-conversion in La2O3:Yb3+,Er3+ microcrystals

with high quantum yield. *Journal of Materials Chemistry C, 5*(42), 11010-11017. doi:10.1039/c6tc05322j

[29] Cho, J., Bass, M., Babu, S., Dowding, J. M., Self, W. T., & Seal, S. (2012). Up conversion luminescence of Yb3+–Er3+ codoped CeO2 nanocrystals with imaging applications. *Journal of Luminescence, 132*(3), 743-749. doi:10.1016/j.jlumin.2011.11.011

[30] Singh, V., Rai, V. K., & Haase, M. (2012). Intense green and red upconversion emission of Er3+,Yb3+ co-doped CaZrO3 obtained by a solution combustion reaction. *Journal of Applied Physics, 112*(6), 063105. doi:10.1063/1.4751450

[31] Singh, V., Rai, V. K., Al-Shamery, K., Haase, M., & Kim, S. H. (2013). NIR to visible frequency upconversion in Er3+ and Yb3+ co-doped BaZrO3 phosphor. *Spectrochimica Acta Part A: Molecular and Biomolecular Spectroscopy, 108*, 141-145. doi:10.1016/j.saa.2013.01.073

[32] Araujo, J. A. (n.d.). Chapter 8. Ultrafine Particles and Atherosclerosis. *Environmental Cardiology,* 198-219. doi:10.1039/9781849732307-00198

[33] Li, D., Wang, C., Strmcnik, D. S., Tripkovic, D. V., Sun, X., Kang, Y., . . . Markovic, N. M. (2014). Functional links between Pt single crystal morphology and nanoparticles with different size and shape: The oxygen reduction reaction case. *Energy Environ. Sci., 7*(12), 4061-4069. doi:10.1039/c4ee01564a

[34] Dong, H., Chen, Y., & Feldmann, C. (2015). Polyol synthesis of nanoparticles: Status and options regarding metals, oxides, chalcogenides, and non-metal elements. *Green Chemistry, 17*(8), 4107-4132. doi:10.1039/c5gc00943j

[35] Verma, A., & Stellacci, F. (2010). Effect of Surface Properties on Nanoparticle-Cell Interactions. *Small, 6*(1), 12-21. doi:10.1002/smll.200901158

[36] Ilawe, N. V., Oviedo, M. B., & Wong, B. M. (2018). Effect of quantum tunneling on the efficiency of excitation energy transfer in plasmonic nanoparticle chain waveguides. *Journal of Materials Chemistry C, 6*(22), 5857-5864. doi:10.1039/c8tc01466c

[37] Rout, K., Mohapatra, M., Layek, S., Dash, A., Verma, H. C., & Anand, S. (2014). The influence of precursors on phase evolution of nano iron oxides/oxyhydroxides: Optical and magnetic properties. *New J. Chem., 38*(8), 3492-3506. doi:10.1039/c4nj00526k

[38] Pola, J., Maryško, M., Vorlíček, V., Bastl, Z., Galíková, A., Vacek, K., . . . Prodan, G. (2005). Infrared laser synthesis and properties of magnetic nano-iron-polyoxocarbosilane composites. *Applied Organometallic Chemistry, 19*(9), 1015-1021. doi:10.1002/aoc.951

[39] Würfel, P., Finkbeiner, S., & Daub, E. (1995). Generalized Planck's radiation law for luminescence via indirect transitions. *Applied Physics A Materials Science & Processing, 60*(1), 67-70. doi:10.1007/bf01577615

[40] Jalil, R. A., & Zhang, Y. (2008). Biocompatibility of silica coated NaYF4 upconversion fluorescent nanocrystals. *Biomaterials, 29*(30), 4122-4128. doi:10.1016/j.biomaterials.2008.07.012

[41] Shan, J., Chen, J., Meng, J., Collins, J., Soboyejo, W., Friedberg, J. S., & Ju, Y. (2008). Biofunctionalization, cytotoxicity, and cell uptake of lanthanide doped hydrophobically ligated NaYF4 upconversion nanophosphors. *Journal of Applied Physics, 104*(9), 094308. doi:10.1063/1.3008028

[42] Boyer, J., & Frank C. J. M. Van Veggel. (2010). Absolute quantum yield measurements of colloidal NaYF4: Er3+, Yb3+ upconverting nanoparticles. *Nanoscale, 2*(8), 1417. doi:10.1039/c0nr00253d

[43] Joubert, M. (1999). Photon avalanche upconversion in rare earth laser materials. *Optical Materials, 11*(2-3), 181-203. doi:10.1016/s0925-3467(98)00043-3

[44] Garnier, N., Moncorgé, R., Manaa, H., Descroix, E., Laporte, P., & Guyot, Y. (1996). Excited-state absorption of Tm3+-doped single crystals at photon-avalanche wavelengths. *Journal of Applied Physics, 79*(8), 4323. doi:10.1063/1.361801

[45] Joubert, M. F., Guy, S., & Jacquier, B. (1993). Model of the photon-avalanche effect. *Physical Review B, 48*(14), 10031-10037. doi:10.1103/physrevb.48.10031

[46] Zou, X., & Izumitani, T. (1993). Spectroscopic properties and mechanisms of excited state absorption and energy transfer upconversion for Er3+-doped glasses. *Journal of Non-Crystalline Solids, 162*(1-2), 68-80. doi:10.1016/0022-3093(93)90742-g

[47] Matsuura, D. (2002). Red, green, and blue upconversion luminescence of trivalent-rare-earth ion-doped Y2O3 nanocrystals. *Applied Physics Letters, 81*(24), 4526-4528. doi:10.1063/1.1527976

[48] Matsuura, D., Ikeuchi, T., & Soga, K. (2008). Upconversion luminescence of colloidal solution of Y2O3 nano-particles doped with trivalent rare-earth ions. *Journal of Luminescence, 128*(8), 1267-1270. doi:10.1016/j.jlumin.2007.12.024

[49] Pang, T., Cao, W., Fu, Y., & Luo, X. (2008). Up-conversion luminescence of trivalent-rare-earth ion-doped LnTaO4 (Ln=Y, Gd, La). *Materials Letters, 62*(16), 2500-2502. doi:10.1016/j.matlet.2007.12.031

[50] Liu, R., Chen, H., Li, Z., Shi, F., & Liu, X. (2016). Extremely deep photopolymerization using upconversion particles as internal lamps. *Polymer Chemistry, 7*(14), 2457-2463. doi:10.1039/c6py00184j

[51] Liang, Z., Cui, Y., Zhao, S., Tian, L., Zhang, J., & Xu, Z. (2014). The enhanced upconversion fluorescence and almost unchanged particle size of β-NaYF4:Yb3+, Er3+ nanoparticles by codoping with K+ ions. *Journal of Alloys and Compounds, 610*, 432-437. doi:10.1016/j.jallcom.2014.04.183

[52] Redmond, S., Rand, S. C., Ruan, X. L., & Kaviany, M. (2004). Multiple scattering and nonlinear thermal emission of Yb3+, Er3+:Y2O3 nanopowders. *Journal of Applied Physics, 95*(8), 4069-4077. doi:10.1063/1.1667274

[53] Mi, C., Wu, J., Yang, Y., Han, B., & Wei, J. (2016). Efficient upconversion luminescence from Ba5Gd8Zn4O21:Yb3+, Er3+ based on a demonstrated cross-relaxation process. *Scientific Reports, 6*(1). doi:10.1038/srep22545

[54] Qu, Y., Kong, X., Sun, Y., Zeng, Q., & Zhang, H. (2009). Effect of excitation power density on the upconversion luminescence of LaF3:Yb3+, Er3+ nanocrystals. *Journal of Alloys and Compounds, 485*(1-2), 493-496. doi:10.1016/j.jallcom.2009.05.145

[55] Wang, J., & Tanner, P. A. (2010). Upconversion for White Light Generation by a Single Compound. *Journal of the American Chemical Society, 132*(3), 947-949. doi:10.1021/ja909254u

[56] Wang, J., Hao, J. H., & Tanner, P. A. (2010). Luminous and tunable white-light upconversion for YAG (Yb_3Al_5O_12) and (Yb,Y)_2O_3 nanopowders. *Optics Letters, 35*(23), 3922. doi:10.1364/ol.35.003922

[57] Strek, W., Marciniak, L., Bednarkiewicz, A., Lukowiak, A., Wiglusz, R., & Hreniak, D. (2011). White emission of lithium ytterbium tetraphosphate nanocrystals. *Optics Express, 19*(15), 14083. doi:10.1364/oe.19.014083

[58] Strek, W., Marciniak, L., Hreniak, D., & Lukowiak, A. (2012). Anti-Stokes bright yellowish emission of NdAlO3 nanocrystals. *Journal of Applied Physics, 111*(2), 024305. doi:10.1063/1.3674272

[59] Strek, W., Marciniak, L., Bednarkiewicz, A., Lukowiak, A., Hreniak, D., & Wiglusz, R. (2011). The effect of pumping power on fluorescence behavior of LiNdP4O12 nanocrystals. *Optical Materials, 33*(7), 1097-1101. doi:10.1016/j.optmat.2010.12.006

[60] Bilir, G., Ozen, G., Collins, J., Cesaria, M., & Bartolo, B. D. (2014). Unconventional Production of Bright White Light Emission by Nd-Doped and Nominally Un-Doped Y2O3 Nano-Powders. *IEEE Photonics Journal, 6*(4), 1-17. doi:10.1109/jphot.2014.2339312

[61] Ogugua, S. N., Ntwaeaborwa, O. M., & Swart, H. C. (2020). Latest development on pulsed laser deposited thin films for advanced luminescence applications. *Coatings, 10*(11), 1078. doi:10.3390/coatings10111078

[62] T. Trupke, E. Daub, P. Würfel. (1998). Absorptivity of silicon solar cells obtained from luminescence. Solar Energy Materials and Solar Cells, Volume 53, Issues 1–2, Pages 103-114, ISSN 0927-0248,

[63] G. Bilir, G. Ozen, M. Bettinelli, F. Piccinelli, M. Cesaria and B. Di Bartolo, "Broadband Visible Light Emission From Nominally Undoped and Doped Garnet Nanopowders," in IEEE Photonics Journal, vol. 6, no. 4, pp. 1-11, Aug. 2014, Art no. 2201211, doi: 10.1109/JPHOT.2014.2337873.

[64] Marciniak, L., Strek, W., Hreniak, D., & Guyot, Y. (2014). Temperature of BROADBAND ANTI-STOKES White emission in LIYBP4O12: Er nanocrystals. *Applied Physics Letters, 105*(17), 173113. doi:10.1063/1.4900965

www.ingramcontent.com/pod-product-compliance
Lightning Source LLC
LaVergne TN
LVHW041449190726
843491LV00008B/2343

* 9 7 8 3 3 8 4 2 6 2 4 6 2 *